Koku in Food Science and Physiology

Toshihide Nishimura • Motonaka Kuroda
Editors

Koku in Food Science and Physiology

Recent Research on a Key Concept in Palatability

Springer

Editors
Toshihide Nishimura
Faculty of Nutrition
Kagawa Nutrition University
Sakado, Chiyoda, Saitama, Japan

Motonaka Kuroda
Institute of Food Sciences & Technologies,
Ajinomoto
Suzuki-cho, Kawasaki, Kanagawa, Japan

ISBN 978-981-13-8455-4 ISBN 978-981-13-8453-0 (eBook)
https://doi.org/10.1007/978-981-13-8453-0

This Springer imprint is published by the registered company Springer Nature Singapore Pte Ltd.
The registered company address is: 152 Beach Road, #21-01/04 Gateway East, Singapore 189721, Singapore

Preface

In Japan, there are lots of foods on whose labels we can see the expression "*Koku* attributes." Why are "*Koku* attributes" used for lots of foods? One reason is that Japanese consumer image palatability of foods from the display of "*Koku* attributes." Japanese use the expression *Koku* in eating some foods. What kind of foods gives us the sensation of *Koku* in Japan? When they eat curry, stew, ramen noodles, cheese, and so on, the expression *Koku* is in general used. On the other hand, when they eat watermelon, Japanese pear, lemon juice, and Japanese apricot, they never use the expression *Koku*. What is different between both groups? The foods with *Koku* attributes give us the sensations such as complexity, mouthfulness, and lingeringness when we put them into our mouth and chew, while other foods without *Koku* attributes do not. These attributes are recognized when they eat foreign foods such as French foods and Chinese foods as well as Japanese foods. However, in these many foods, it is still unclear what the sensation of *Koku* is.

Several years ago, we organized "*Koku* Attribute Association" and started discussing about the definition of "*Koku* attribute." At first, we proposed that the sensation of "*Koku* attributes" is caused by lots of stimulations of taste, aroma, and texture. Furthermore, we made objective elements such as complexity, mouthfulness and lingeringness in the factor "*Koku* attributes," as five basic tastes make up taste.

Complexity is essential for foods with "*Koku* attributes," because lots of stimulations of taste, aroma, and texture are given to us in these foods and give the characteristic of the foods with "*Koku* attributes." These lots of stimulations are produced during postmortem aging, heating, and fermentation in the production of foods. Mouthfulness and lingeringness are brought by umami compounds and lipids. In the presence of umami compounds and/or lipids, many stimulations spread into oral cavity and linger in our mouth. The compounds such as *kokumi* substances, aroma compounds, and maillard peptides have already been shown to enhance mouthfulness in *Koku* attribute of foods. We think that three elements in *Koku* attributes are different in the appropriate intensity depending on foods. For example, there are the expressions such as weak, medium, and strong. Depending on the amount of umami compounds, lipids and *kokumi* compounds and so on, the *Koku* intensity should be able to be changed from weak to strong. And *Koku* attributes are possible to be objectively measured.

However, the research on *Koku* attributes have just started. Having this in mind, in this book, we asked to write manuscript of chapters to authors who are doing research on *Koku* attributes. Chapter 1 provides the definition of "*Koku* attributes." *Koku* is the unique sensation caused by lots of stimulations of taste, aroma, and texture and is constituted by three elements, that is, complexity, mouthfulness, and lingeringness. Furthermore, the compounds responsible for *Koku* enhancer are also introduced. It will help the readers to understand what "*Koku* attribute is." Chapters 2 and 3 provide accurate functions of glutamate in terms of its taste and its flavor-enhancing action. Umami compound is also shown to be important in the *Koku* attribute by enhancing continuity and mouthfulness of the overall taste experience. Chapters 4 and 5 provide compounds responsible for the formation and the enhancement of *Koku* attributes in pork sausage and beer. Umami compound and fats enhance mouthfulness and lingeringness of *Koku* attributes in pork sausages. Aroma compounds are contributing to the complexity or thickness of *Koku* attribute in beer aroma. Chapter 6 provides odor compounds which enhance *Koku* attribute of foods. They are also shown to be a phthalide compound in celery, (4Z, 7Z)-trideca-4,7-dienal in dried bonito, and rotundone in several fruits. Chapters 7, 8 and 9 provide the *kokumi* substances and their functions on the sensory characteristics of foods and their mechanisms for the enhancement of *Koku* attributes from physiology aspect. γ-glutamyl-valyl-glycine and glutathione are shown to be *kokumi* peptides and enhance oiliness, thick flavor, and aftertaste of foods. These peptides are clarified to be an agonist of calcium-sensing receptor (CaSR) and stimulate the mouse trigeminal neurons. Chapter 10 provides the overview of the study on *Koku*.

I hope that the information provided in this book shall give an outlook of the definition of "*Koku* attributes," the compounds involved in *Koku* attributes of some foods, and the physiology concerning with the sensation of "*Koku* attributes" to the novices in the field, as well as useful information for more experienced readers, and can be a concise starting book for all interested in food science and physiology of "*Koku* attributes."

Sakado, Chiyoda, Saitama, Japan — Toshihide Nishimura
Suzuki-cho, Kawasaki, Kanagawa, Japan — Motonaka Kuroda

Contents

Chapter 1
Definition of "*Koku*" Involved in Food Palatability

Toshihide Nishimura

Abstract It is well known that many factors—taste, aroma, tenderness, juiciness, viscosity, and so on—are involved in food palatability. In Japan, the term "*koku*" (referring to *koku* attributes) is used in eating and evaluating to be palatable foods such as curry, stew, ramen noodles, natural cheese, and so on. Thus, it has been thought to be one of the important factors enhancing food palatability. However, *koku* has not yet been defined, and there have been few research studies on the compounds that enhance *koku* (*koku* attributes).

This chapter introduces definitions of *koku* and *koku* attributes. *Koku* is the unique sensation caused by lots of stimulation in terms of taste, aroma, and texture, which is expressed as complexity in *koku*. This is the essence of *koku* attributes. Furthermore, when this sensation is spread out without changing its characteristic flavor in our whole oral cavity, and it lingers after swallowing, we can sense strong *koku* attributes such as mouthfulness and lingeringness. Therefore, *koku* is constituted by three elements: complexity, mouthfulness, and lingeringness (continuity). Complexity in the *koku* attributes of foods is produced by aging, fermentation, and heating, causing production of various compounds. Mouthfulness and lingeringness (continuity) have recently been explained as being caused by umami compounds and/or phytosterols in onions.

Recent sensory evaluation analyses have shown that the addition of umami compounds enhances the retronasal aroma sensation, resulting in enhancement of the flavor of foods (mouthfulness). Umami compounds have also been shown to cause an effect of lingeringness of the taste sensation. Furthermore, phytosterols in onion have been confirmed to contribute to a sensation of lingeringness of the aroma—a *koku* attribute. Headspace gas chromatography analyses have shown that β-sitosterol binds with methyl propyl disulfide and hexanal. The addition of 0.02% β-sitosterol to consommé soup enhances the mouthfulness and lingeringness of the aroma sen-

---

T. Nishimura (✉)
Faculty of Nutrition, Kagawa Nutrition University, Sakado-city, Saitama, Japan
e-mail: nishimura.toshihide@eiyo.ac.jp

T. Nishimura, M. Kuroda (eds.), *Koku in Food Science and Physiology*,
https://doi.org/10.1007/978-981-13-8453-0_1

sation. On the basis of these findings, umami compounds and phytosterols are thought to be candidates for enhancement of *koku* attributes in palatable foods.

Keywords *Koku* · Complexity · Mouthfulness · Lingeringness · Umami compounds · Lipids

1.1 *Koku* Attributes in Foods

For a long time, in Japan, the term "*koku*" (referring to *koku* attributes) has been used to describe palatable foods, such as when we eat curry and, stew, ramen noodles, natural cheese, and so on. Recently, the term "*koku*" has begun appearing on the packaging of many foods such as mayonnaise, coffee, cocoa, yogurt, pudding, *kimuchi*, beer, and soy sauce. Since they are accustomed to saying that there are *koku* attributes in the foods they find palatable, the use of the term "*koku*" on food packaging makes many Japanese people think of palatable foods (Fig. 1.1). However, we recognize that "*koku*" is not necessarily a synonym for the terms "palatability" and "deliciousness." Although pear, watermelon, lemonade, and umeboshi (pickled ume) are considered palatable foods by many Japanese people, they never say that there are *koku* attributes in these foods (Fig. 1.2). These foods are not heated, conditioned (i.e., aged), or fermented, so there is a simple sensation but no sensations of complexity, mouthfulness, and lingeringness. On the other hand, foods such as curry, stew, ramen noodles, and natural cheese—which are heated, conditioned, or fermented—give us sensations such as complexity, mouthfulness, and lingeringness, and when people find the complexity, mouthfulness, and lingeringness of these foods suitable, they consider these foods palatable with *koku* attributes.

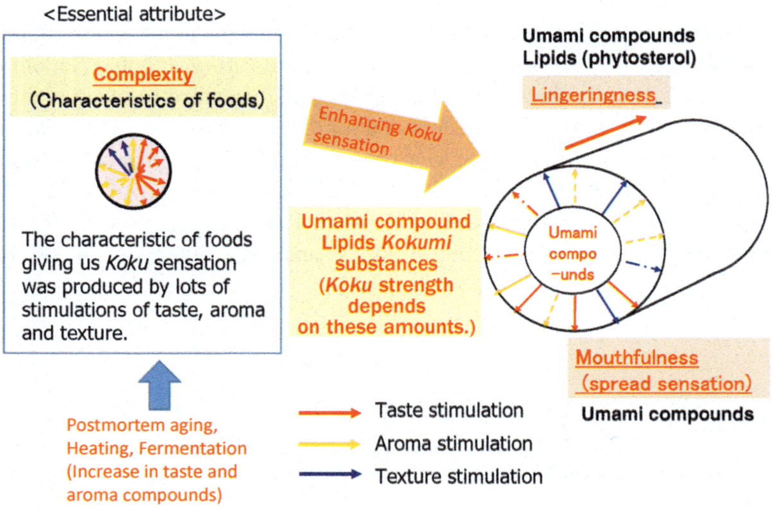

Fig. 1.1 The definition of *koku*

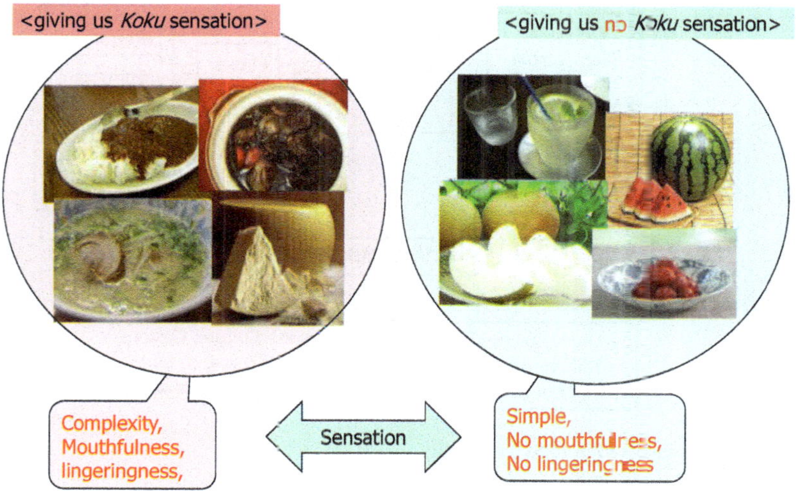

Fig. 1.2 What kind of foods convey *koku*?

However, when people find these characteristics unsuitable in a food, they consider them unpalatable because they contain too much or too little complexity, mouthfulness, and/or lingeringness. Therefore, the term "*koku*" is not a synonym for the terms "palatability" and "deliciousness." It is also thought that palatability and deliciousness are determined by subjective evaluation, whereas *koku* attributes are determined by objective evaluation. This is the first time that this concept has been proposed, because *koku* attributes have not yet been defined, and there have been few research studies on the compounds that enhance *koku* (*koku* attributes).

1.2 Definitions of *Koku* and *Koku* Attributes

1.2.1 Factors Involved in Food Palatability

Many factors—such as taste, aroma, texture (tenderness, viscosity, smoothness, juiciness, and so on), color, temperature, and shape—are involved in food palatability (Fig. 1.3). Among many factors, in general, taste is caused by taste compounds in foods, which are water soluble. Aroma is caused by aroma compounds, which are volatile compounds released from foods. Aroma is classified into orthonasal and retronasal aromas, and the latter aroma is sensed after foods are put into the oral cavity. This retronasal aroma is thought to be the most important factor involved in food palatability. Recently, an interaction between taste and retronasal aroma has been described. Previously, *koku* attributes were not classified into objective factors involved in food palatability, because the meaning of *koku* (*koku* attributes) was thought to be the same as palatability. However, it is now thought that *koku* is one of

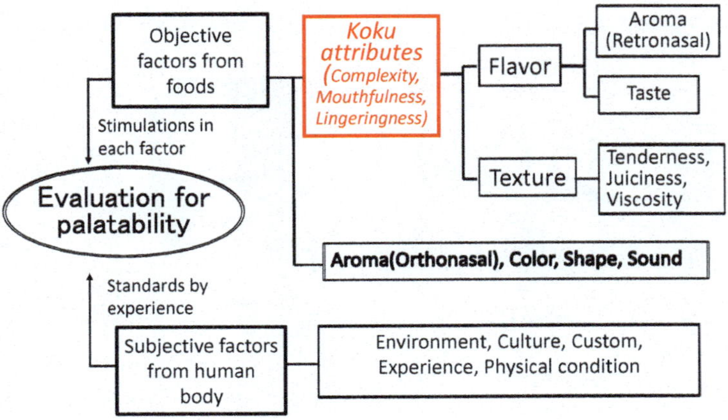

Fig. 1.3 Factors involved in food palatability

the objective factors involved in food palatability, and the sensations of complexity, mouthfulness, and lingeringness in the *koku* attributes of foods such as curry, stew, ramen noodles, and natural cheese are thought to be caused by the entirety of factors such as taste, aroma, and texture.

1.2.2 Definition of Koku *Attributes*

Given the background mentioned above, foods such as curry, stew, ramen noodles, and natural cheese—which are heated, conditioned, or fermented for a long time—possess *koku* attributes such as complexity, mouthfulness, and lingeringness (Fig. 1.1). It has been confirmed that *koku* attributes are mainly conveyed by a lot of stimulation from taste compounds, aroma compounds, and texture. Many consumers believe that palatability is caused only by the taste of a food after it is put into their oral cavity. However, when we eat a food while pinching our nose, we cannot identify what kind of food it is and do not evaluate that food as being palatable. This means that retronasal aromas play an important role in our evaluation of the palatability of foods. We also have a similar experience of finding the foods we eat unpalatable when we have a blocked nose because of catching a cold. This is the reason why we cannot sense *koku* attributes and the characteristic aromas of foods. Therefore, the number and the intensity of stimulations from taste and aroma compounds, and from texture, convey the intensity of the complexity of *koku* attributes.

There is an objective intensity in the *koku* attributes of complexity, mouthfulness, and lingeringness, and this intensity depends on the amount of stimulation from taste compounds, aroma compounds, and the compounds involved in texture. The compounds that contribute to *koku* attributes and the intensity of *koku* attributes differ depending on the food. Research on these compounds and the intensity of *koku* attributes in each food is needed.

1.3 Elements and Compounds Involved in *Koku* Attributes

When we eat foods with *koku* attributes, we sense their complexity, mouthfulness, and lingeringness. So, I propose that these three sensations are basic elements of *koku* attributes, just as the five basic elements of taste are sweetness, saltiness, sourness, bitterness, and umami (Fig. 1.4).

1.3.1 Complexity

The complexity of foods with *koku* attributes is harmoniously formed by lots of stimulation from taste compounds, aroma compounds, and texture (Fig. 1.5). When we get lots of stimulation unharmoniously from foods, we do not sense complexity. Therefore, harmony already exists in complexity and is not a basic element of *koku* attributes.

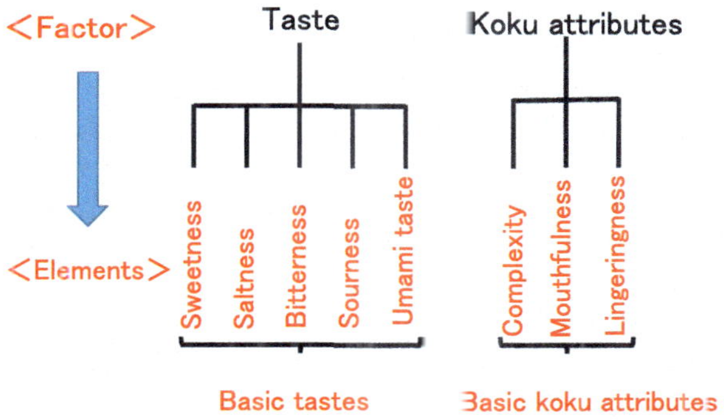

Fig. 1.4 Elements of taste and *koku* attributes

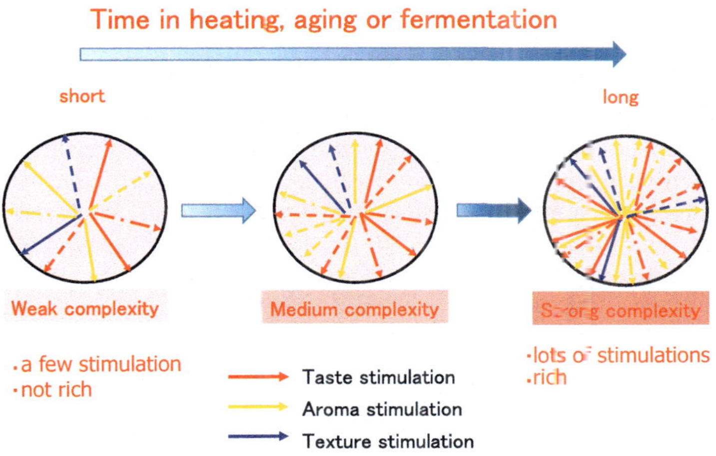

Fig. 1.5 Formation of complexity in *koku* attributes of foods

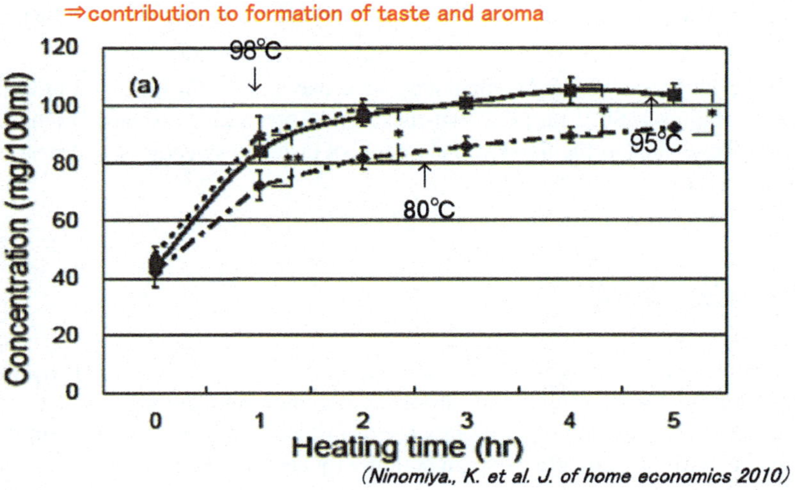

Fig. 1.6 Changes in free amino acids extracted from materials during heating

The stimulation in foods with *koku* attributes is produced by heating, fermentation, and conditioning for a long time. Stew and curry are cooked by heating for a long time. During heating, various taste compounds are extracted from foodstuffs and various aroma compounds are produced by Maillard reactions between amino acids and sugars (Fig. 1.6). Miso paste and soy sauce are produced by fermentation with microorganisms such as *Aspergillus* and lactic acid bacteria. These microorganisms produce free amino acids and peptides by protein degradation and aroma compounds by enzyme reactions and by Maillard reactions (Fig. 1.7). Natural cheeses are also conditioned for long time. During conditioning of natural cheeses, taste compounds such as free amino acids and peptides are produced through protein degradation by proteases in microorganisms. Many kinds of aroma compounds are also produced from lipids by microorganisms. In particular, the types of aroma compounds differ between natural cheeses because of the difference in the microorganisms in those cheeses.

The constituents of taste and aroma compounds involved in complexity lead to the characteristics of these foods. The intensity of the complexity in these foods can be varied by changing the duration of heating, conditioning, or fermentation, leading to differences in the flavor characteristics of these foods. We prepared a synthetic chicken extract using compounds that had been listed as constituents of the taste of chicken in a previous paper. When we omitted one free amino acid, the intensity of the retronasal aroma was found to decrease.

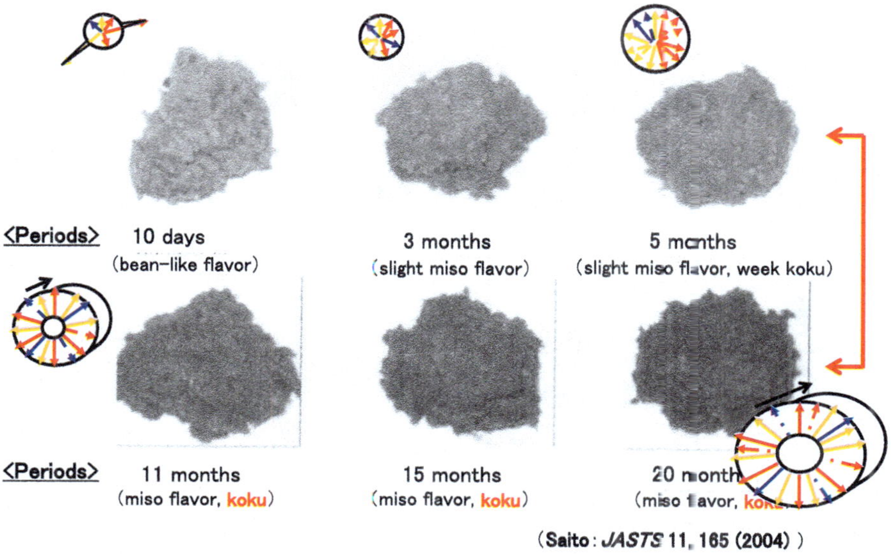

Fig. 1.7 Increase in the intensity of *koku* attributes during fermentation of miso paste

1.3.2 Mouthfulness

The mouthfulness in the *koku* attributes of foods such as curry and stew is sensed by the spread of the flavor sensation in the space of our oral cavity without changing its characteristic flavor. This spread of the flavor sensation is thought to be caused by umami substances. A miso soup concentration of 50 g/L without seasonings provides a very weak sensation in the space of our oral cavity, although it has the characteristic flavor and complexity of miso soup. The addition of umami substances to this miso soup can increase the intensity of the spread of the flavor sensation in our oral cavity without changing its characteristic flavor. Yamaguchi and Kimizuka (1979) reported that umami substances had the effect of flavor enhancers when they were added to foods. Recently, Nishimura et al. (2016a) examined the mechanism of flavor enhancement with addition of umami substances to foods. Their analysis, using a model chicken extract, clarified that addition of monosodium glutamate (MSG) to the extract increased the sensation of the retronasal aroma by 2.5 times (Fig. 1.8). This effect increased when the concentration of added MSG was increased up to a value of 0.3%. However, addition of MSG in a concentration over 0.3% made the sensation of the retronasal aroma of the extract weaker, while the intensity of the umami taste became stronger. Therefore, umami substances have the strong effect of conveying the mouthfulness *koku* attribute by enhancing flavor, especially the retronasal aroma.

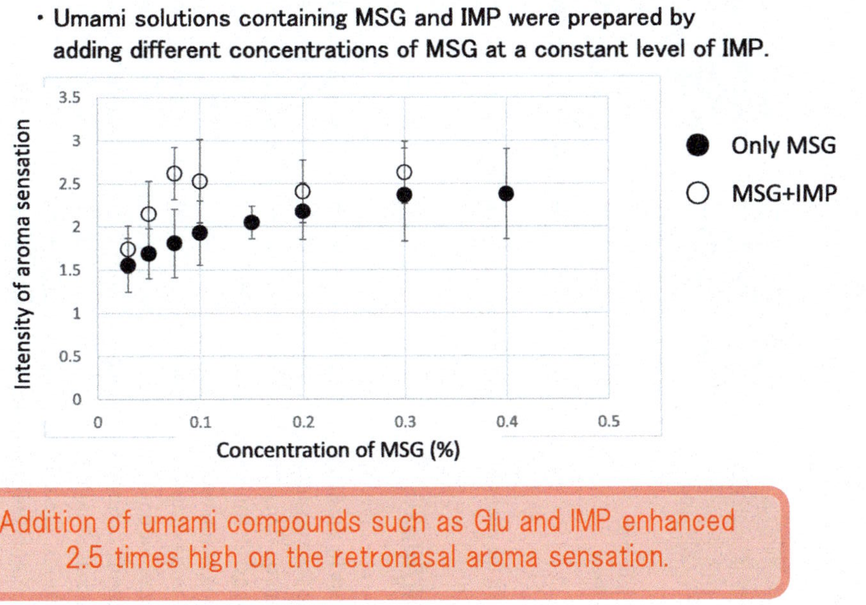

Fig. 1.8 Effects of umami compounds on the intensity of the aroma sensation, using a flavored solution

There have been some research studies on compounds involved in the mouthfulness *koku* attribute. Ueda et al. (1990) found that addition of an aqueous extract of garlic to soups enhanced the sensations of continuity, mouthfulness, and thickness in the soups. This extract was also found to have this enhancing effect when it was added to an umami solution composed of 0.05% MSG and disodium inosinate. The key compounds in the aqueous extract of garlic were clarified to be alliin, S-methyl-L-cysteine sulfoxide, and γ-L-glutamyl-S-allyl-L-cysteine. Furthermore, the addition of alliin at 0.05% (w/v), which had no aroma or taste, was reported to have this enhancing effect on the soups. Ueda et al. (1994) also clarified that trans-S-propenyl-L-cysteine sulfoxide (PeCSO) or its γ-glutamyl peptide (γ-Glu-PeCSO) enhanced the sensations of continuity, mouthfulness, and thickness in an umami solution when added in a concentration of 0.02% (w/v), although these compounds at this concentration did not add any aroma or taste to the umami solution.

Dunkel and Hofmann (2009) reported that addition of a nearly tasteless aqueous extract isolated from beans to a model chicken broth enhanced the sensations of mouthfulness and complexity, and successively induced a long-lasting savory taste on the tongue. They clarified that γ-L-glutamyl-L-leucine, γ-L-glutamyl-L-leucine, and γ-L-glutamyl-L-cysteinyl-β-alanine were the key molecules, and they called them *kokumi* peptides. Then, they clarified that γ-L-glutamyl peptides such as γ-glutamyl-Glu and γ-glutamyl-Gly, found in Gouda cheese, were key compounds that enhanced the mouthfulness and lingeringness of the matured cheese. Kuroda et al. (2013) discovered the *kokumi* peptide γ-L-glutamyl-L-valyl-glycine (γ-EVG), which is found in soy sauce, raw scallop, and processed scallop products. These *kokumi* peptides are perceived by calcium-sensing receptors (CaSRs) (Ohtsu et al. 2010). However, there has been no clarification of the mechanisms involved in the sensation of mouthfulness and lingeringness conveyed by the binding of *kokumi* peptides to CaSRs.

Ogasawara et al. (2006a, b) found that Maillard peptides with a molecular weight of 1000–5000 Da from soybean paste enhanced the sensations of lingeringness and mouthfulness in an umami solution and consommé soup when these peptides were added to these solutions. Maillard peptides, which are increased during fermentation in the production process of aged miso, were thought to be key substances in enhancing the characteristic flavor of miso.

Kurobayashi et al. (2008) reported that addition of three phthalides—sedanenolide, 3-n-butylphthalide, and sedanolide—to chicken broth enhanced the umami intensity, as well as the complexity, of the broth when these phthalides were added to the broth at the lower concentration than threshold level. Thus, although some substances have been discovered to possess the enhancing effect of mouthfulness, the enhancing mechanisms by these compounds were not clarified.

1.3.3 Lingeringness (Continuity)

The lingeringness in *koku* attribute of foods such as curry and stew is sensed in the whole oral cavity and in the nose as a retronasal aroma.

The mechanism accounting for the lingeringness of umami substances sensed in the oral cavity and the mixture of umami substances and *kokumi* peptides or Maillard peptides are still not clearly understood. When an umami compound is put into the mouth, the tactile stimulation of the tongue continues for a long time. Although sulfur-containing compounds or *kokumi* peptides also show the effect of lingeringness, there is little evidence that only these substances can show the same effect that umami compounds do.

Another example of the lingeringness *koku* attribute is caused by the binding activity of aroma compounds with lipids. Ramen noodles in pork bone soup and marbled beef from Japanese black cattle are rich dishes because of the lipid or fat content in these foods. Although pure lipids or fats have no taste and aroma in general, we can sense a complex taste and aroma when we put cooked lipids or fats into our mouth. Nishimura et al. (2016b) clarified that this phenomenon is brought about by taste and aroma compounds attached to the lipids and fats (Fig. 1.9). They found that a precipitate of heat-treated onion concentrate (HOC) interacted with aroma compounds and thus enhanced aroma persistence, including the lingeringness of the aroma, as a *koku* attribute of consommé. The key compounds in the precipitate were phytosterols—that is, β-sitosterol and stigmasterol. To my knowledge, this was the first study to show that phytosterols can interact with aroma compounds in foods. Lipids or fats in foods contribute to complexity and lingeringness as *koku* attributes

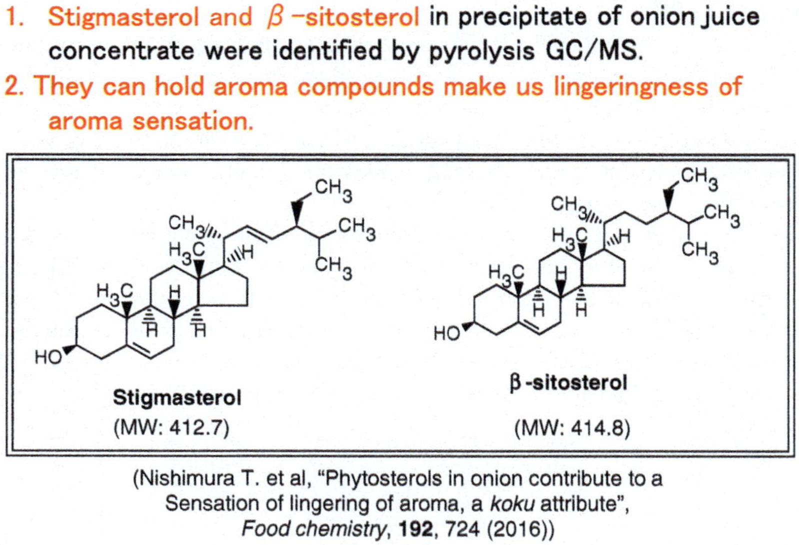

Fig. 1.9 Compounds in precipitate of onion juice concentrate

of foods. It is well known that the presence of lipids influences the persistence of aroma compounds in foods. Aroma compounds in foods partially bind to lipids or fats by hydrophobic interaction.

1.4 Regulation of the Intensity of *Koku* Attributes in Food Development

To develop new food products possessing *koku* attributes, first an image of the sensations of foods should be created (Fig. 1.10). For example, we may want to produce foods with weak lingeringness and strong complexity and mouthfulness, or foods with strong complexity and lingeringness. To regulate the intensity of *koku* attributes, the methods described in Sects. 1.4.1, 1.4.2, and 1.4.3 seem to be effective.

1.4.1 Production of Foods Possessing Koku *Attributes*

As mentioned earlier, the complexity of *koku* attributes in foods is mainly produced by heating, fermentation, or conditioning. The intensity of the complexity in *Koku* attributes of these foods is regulated by the duration of these production processes.

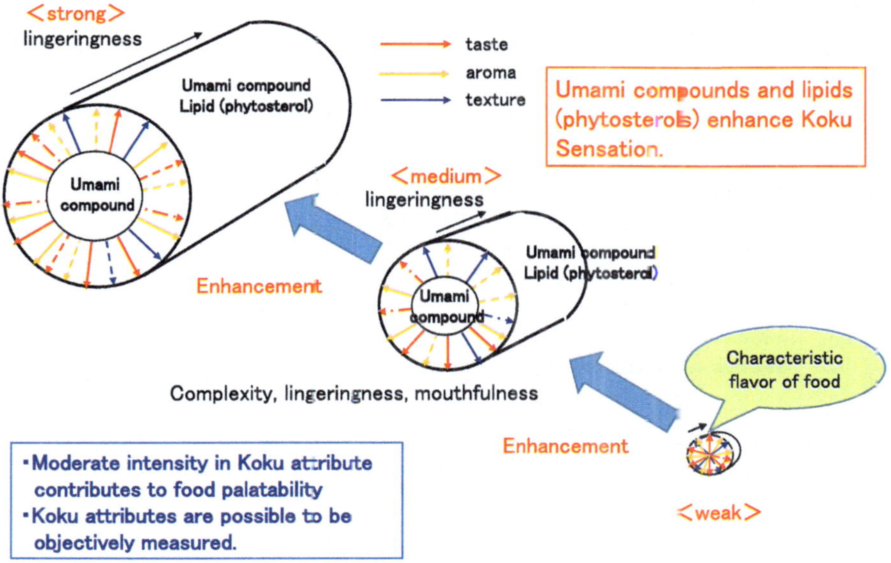

Fig. 1.10 Intensity of *koku* attributes

Bouillon is prepared by heating meats and various vegetables—such as carrot and onion—in soup for long time. During this heating process, free amino acids, sugars, and organic acids are extracted from the foodstuffs, and various favorable aroma compounds are also produced in the soup by the Maillard reaction between amino acids and sugars. The intensity of the complexity of the *koku* attributes of this bouillon is regulated by changing the duration of heating. In general, the complexity of these foods becomes stronger when they are heated for a long time rather than for a short time, because the extraction of taste compounds with heating increases with time.

Natural cheeses and some meats are conditioned for a long time at a certain temperature to produce favorable flavors in these foods. During conditioning of natural cheeses and meats, many kinds of taste compounds, such as free amino acids and peptides, are produced by endogenous or microorganism proteases, which enhance the complexity of the *koku* attributes of these foods. In the case of cheeses, specific aroma compounds are produced from fatty acids by the action of specific microorganisms. This complexity becomes stronger when they are conditioned for a longer time. In the case of meats, meaty and specific aromas of meat are produced by a Maillard reaction when the meats are heated. Umami compounds also increase during conditioning, leading to enhanced mouthfulness and lingeringness (Fig. 1.11).

Miso paste is produced by fermentation using soybeans. As the fermentation of the miso paste proceeds, the color of the paste changes from yellow to brown because of the Maillard reaction. During fermentation of miso paste, various kinds

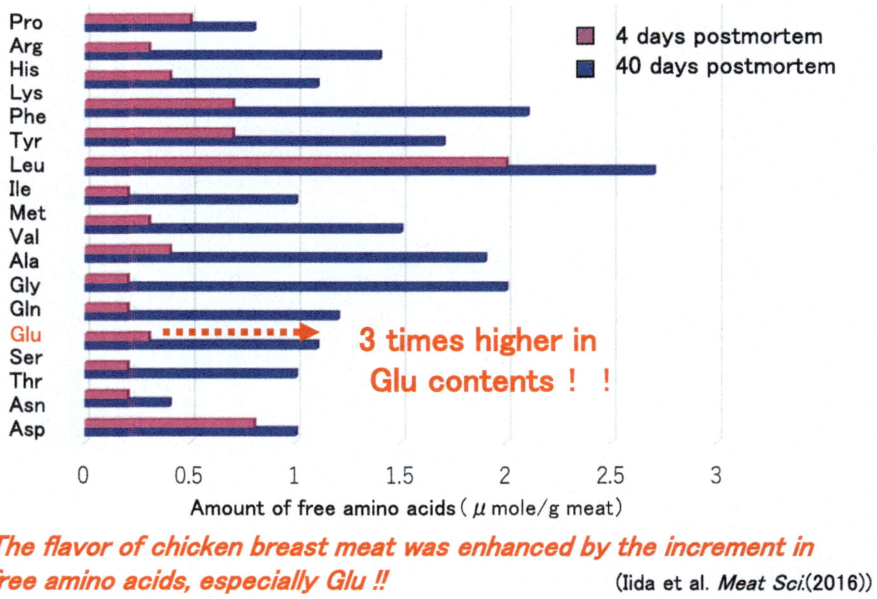

Fig. 1.11 Increase in free amino acids during postmortem conditioning of the loin of Japanese black cattle

of taste compounds—such as free amino acids, peptides, and aroma compounds—are produced and increased by the action of microorganisms. The incrementation of taste and aroma compounds causes complexity in the *koku* attributes of miso paste. Furthermore, incrementation of umami compounds and Maillard peptides during fermentation also occurs and enhances the mouthfulness and lingeringness of miso paste. Thus, lengthy fermentation of miso paste produces stronger *koku* attributes than brief fermentation.

1.4.2 Application of Substances with Enhancing Effects on Koku *Attributes*

When we increase the intensity of mouthfulness and lingeringness as *koku* attributes of foods, umami substances, both umami substances and *kokumi* substances, or both umami substances and Maillard peptides should be added to foods that have weak *koku* attributes. We examined the effect of the amount of umami compounds added to pork sausages on the *koku* attributes of this food. The addition of umami substances to sausages enhanced the intensity of the sensation of a retronasal aroma, the complexity of the sausage flavor, and the lingeringness of the taste.

We clarified that the addition of lipids to foods can enhance the intensity of the *koku* attributes of foods. First, β-sitosterol, which has the ability to bind to aroma compounds was added to Chinese soup, and the intensity of the soup's spicy aroma was enhanced (Figs. 1.12 and 1.13). The intensity of the lingeringness of the aroma and the complexity and mouthfulness of the soup were also enhanced. We also clarified that the addition of fat to pork sausages enhanced the intensity of the mouthfulness and complexity of the sausages.

1.4.3 Combinations with Foods Possessing Koku *Attributes*

In Japan there are delicious foods such as sashimi (sliced raw fish), *basashi* (raw horse meat), and *moro-kyu* (cucumber with *moromi* miso paste). Raw fish, raw horse meat, and cucumber are foods with no *koku* attributes. However, we eat raw fish dipped in soy sauce, raw horse meat dipped in *tare* sauce (a special sauce), and raw cucumber with miso paste put on its surface. Soy sauce, *tare* sauce, and miso paste are foods with *koku* attributes. Even if there are no *koku* attribute in foods, we can change these foods into foods that do possess *koku* attributes by combining them with other food materials such as soy sauce and miso paste.

Japanese soup stock (dashi) is made with dried seaweed (kombu) and bonito shavings, and it has weak *koku* attributes because it contains umami substances. Some Japanese chefs say that the addition of small dried sardines and mackerel shavings to dashi can make its *koku* attributes much stronger.

Thus, we can regulate the intensity of the *koku* attributes of foods by combination with food materials that have stronger *koku* attributes.

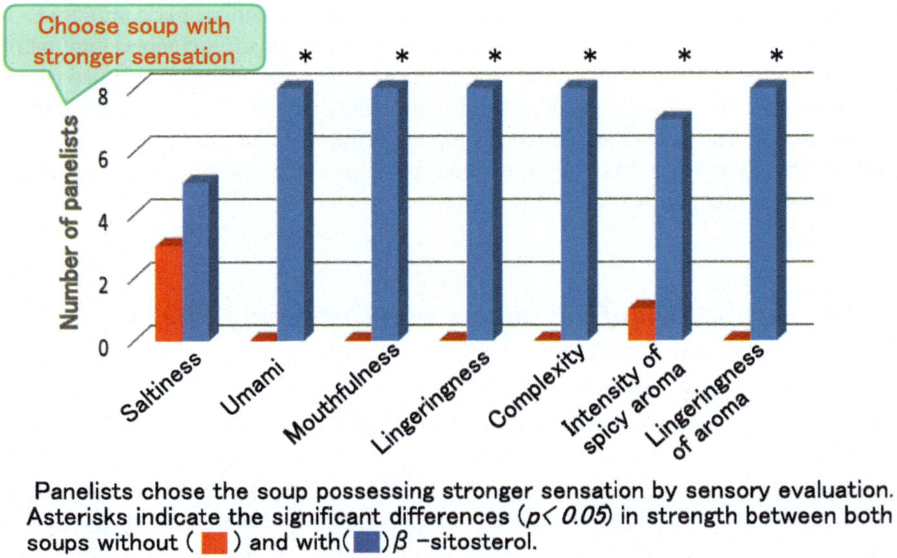

Fig. 1.12 Effect of addition of 0.05% β-sitosterol on the sensory characteristics of Chinese soup

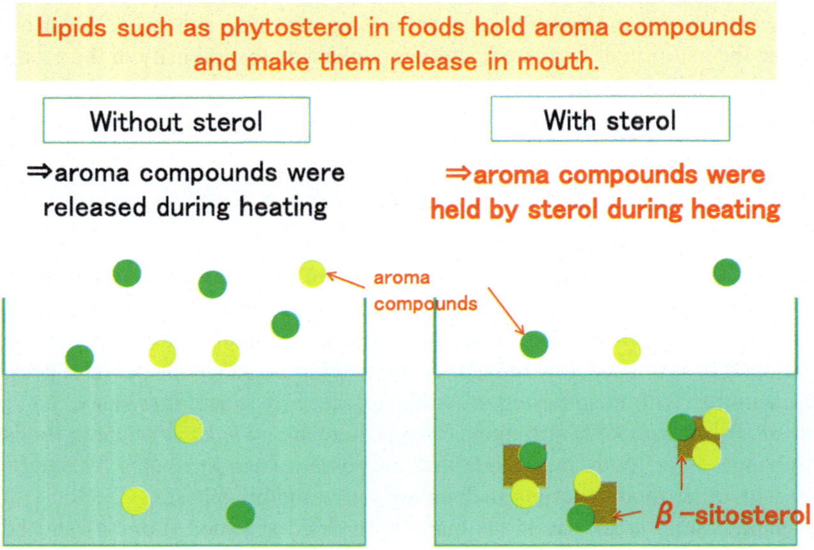

Fig. 1.13 Mechanism of the lingeringness of aromas, caused by phytosterols

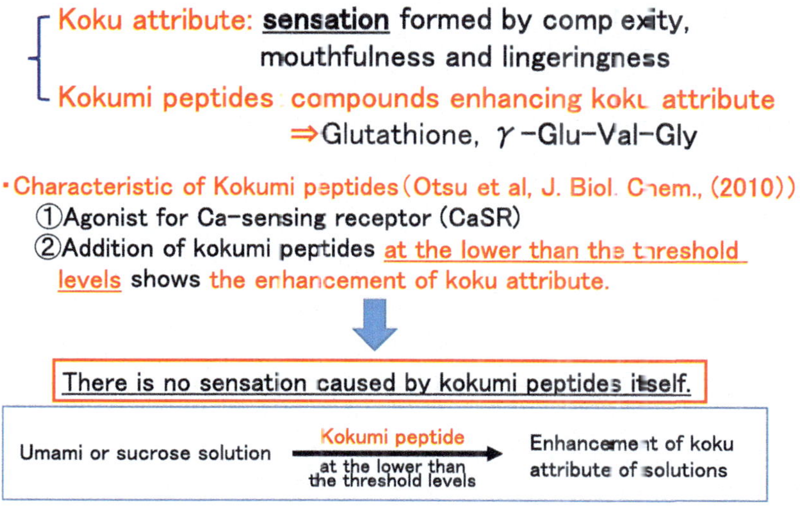

Fig. 1.14 Differences in *koku* attributes and *kokumi* peptides

1.5 *Koku* Differs from *Kokumi* in Food Sensations (Fig. 1.14)

Koku is quite different from *kokumi*, because of the difference in these sensations. *Koku* is the whole sensation caused by taste, aroma, and texture, and it is composed of three elements: complexity, mouthfulness, and lingeringness. On the other hand *kokumi* was initially thought to be the sixth taste, because *kokumi* substances bind to CaSRs. However, *kokumi* substances such as alliin, PeCSO, glutathione, and γ-EVG enhance the sensations of mouthfulness and lingeringness at the lower concentrations than their threshold levels, and they can cause to convey these sensations of soup or a umami solution only in the presence of umami substances. Therefore, there is no specific expression for *kokumi*. It should be noted that alliin, PeCSO, glutathione, γ-EVG, and so on are *kokumi* substances, and the addition of these substances to soup enhances the mouthfulness and lingeringness *koku* attributes of the soup.

References

Dunkel A, Hofmann T (2009) Sensory-directed identification of b-alanyl dipeptides as contributors to the thick-sour and white-meaty orosensation induced by chicken broth. J Agric Food Chem 57:9867–9877

Kurobayashi Y, Katsumi Y, Fujita A, Morimitsu Y, Kubota K (2008) Flavor enhancement of chicken broth from boiled celery constituents. J Agric Biol Chem 56:512–516

Kuroda M, Kato Y, Yamazaki J, Kageyama N, Mizukoshi T, Miyama H, Eto Y (2013) Determination of γ-glutamyl-valyl-glycine in scallop and processed scallop products using high pressure liquid chromatography–tandem mass spectrometry. Food Chem 141:823–828

Nishimura T, Goto S, Miura K, Takakura Y, Egusa AS, Wakabayashi H (2016a) Umami compounds enhance the intensity of reteronasal sensation of aromas from model chicken soups. Food Chem 196:577–583

Nishimura T, Egusa AS, Nagao A, Odahara T, Sugise T, Mizoguchi N, Nosho Y (2016b) Phytosterols in onion contribute to a sensation of lingering of aroma, a *koku* attribute. Food Chem 192:724–728

Ogasawara M, Katsumata E, Egi M (2006a) Taste properties of Maillard-reaction products prepared from 1000 to 5000 Da peptide. Food Chem 99:600–604

Ogasawara M, Yamada Y, Egi M (2006b) Taste enhancer from the long-term ripening of miso (soybean paste). Food Chem 99:736–741

Ohtsu T, Amino Y, Nagasaki H, Yamanaka T, Takeshita S, Hatanaka T, Maruyama Y, Miyamura N, Eto Y (2010) Involvement of the calcium-sensing receptor in human taste perception. J Biol Chem 285:1016–1022

Ueda Y, Sakaguchi M, Hirayama K, Mijajima R, Kimizuka A (1990) Characteristic flavor constituents in water extract of garlic. Agric Biol Chem 54:163–169

Ueda Y, Tsubuku T, Miyajima R (1994) Composition of sulfur-containing components in onion and their flavor characters. Biosci Biotechnol Biochem 58:108–110

Yamaguchi S, Kimizuka A (1979) Psychometric studies on the taste of monosodium glutamate. In: Filer LJ Jr et al (eds) Glutamic acid: advances biochemistry and physiology. Raven Press, New York, pp 35–54

Chapter 2
Umami and *Koku*: Essential Roles in Enhancing Palatability of Food

Takashi Yamamoto

Abstract "Umami" (or umami taste) and "*koku*" are Japanese words born and used for a long time in the characteristics of traditional Japanese cuisine, and both are related to deliciousness of food mainly on the basis of functions of glutamate. Although umami is now worldwidely accepted as a basic taste together with the four basic tastes, sweet, salty, sour, and bitter, it is needed to know the accurate functions of glutamate in terms of its taste, umami, and flavor-enhancing action. Umami, from its discovery to its receptor mechanisms, is described for a better understanding of its taste. The flavor-enhancing action is related to *koku*. *Koku* is a conceptual word used by the Japanese on a daily basis for foods and drinks, implying enhanced deliciousness accompanying "thickness," "continuity," and "mouthfullness" in flavors and textures. *Koku* could be induced by integration of odor, texture, and taste of complex ingredients in foods. Concerning only the taste aspect of *koku,* which is called *"kokumi,"* it is supposed that responses to umami substances are enhanced by certain substances (*kokumi*-inducing substances) contained in complex ingredients or added from outside. *Kokumi* substances are known to be agonists for calcium-sensing receptor (CaSR). Better and accurate knowledge on umami and *koku* is essential to understand the important roles of glutamate in inducing palatability of food.

Keywords Glutamate · Flavor-enhancing action · Thickness · Continuity · Mouthfullness · *Kokumi* substances · CaSR

T. Yamamoto (✉)
Department of Nutrition, Faculty of Health Science, Kio University, Nara, Japan
e-mail: ta.yamamoto@kio.ac.jp

T. Nishimura, M. Kuroda (eds.), *Koku in Food Science and Physiology*,
https://doi.org/10.1007/978-981-13-8453-0_2

2.1 Characteristics of Japanese Cuisine

2.1.1 Rich Variety of Ingredients

Japanese food is made of a great variety of ingredients, reflecting the geographical and climatic circumstances of the Japanese islands. Japan is surrounded by sea with a convergence of warm and cold currents, which provides an ideal environment for a variety of marine life such as fish, shellfish, and seaweed. The Japanese archipelago stretches for over 3000 km, from the cold northernmost island, Hokkaido, through to the subtropical clime of Okinawa. Together with its length, hills and mountains occupying a large part of the land give a wide variety of fruit, vegetables, nuts, and fungi. Moreover, a temperate climate of four different seasons, spring, summer, autumn, and winter, results in a wide array of delicious natural ingredients. The flat areas of Japan have traditionally been used to grow rice as the staple crop since the techniques of paddy field cultivation were introduced from China.

2.1.2 Dashi *Culture*

Japanese eat rice with dishes of various ingredients for their daily meals. Although Japanese basically enjoy the natural tastes and textures of the ingredients, they are traditionally cooked with "*dashi*" (Japanese for soup stock) to give them appetitive flavors (tastes and odors). *Dashi* is an important ingredient which has been used in Japanese cuisine for more than 1000 years. Thus, *dashi* can be said as a base of traditional Japanese food. Examples of *dashi* include all-purpose soup stock that is usually made from "*kombu*" (or "*konbu*," dried seaweed), "*katsuobushi*" (dried bonito), and dried "*shiitake*" mushroom. *Dashi*, which literally means "boiled extract," has a very simple and subtle taste compared with the complex taste of soup stock in China and Western countries. *Dashi* is used to add good flavors to almost all boiled dishes and soups, just as bouillon is used in Western or Chinese cooking.

Kombu is a type of sea vegetable, known as kelp (brown kelp, tangle, genus Laminaria), and grows in cold seawater mainly around Hokkaido island in Japan. It takes about 2 years for the *kombu* to be fully grown. As soon as it is harvested, the *kombu* is dried in the sun and then by hot-air fans in drying chambers. The dried *kombu* is then allowed to mature for 2 years, during which time its flavor develops further. Preparing "*kombu-dashi*," a soup stock prepared with *kombu*, is very simple: cut a dried kombu sheet into about 20 cm, heat the *kombu* and 1 liter of water in a large saucepan over medium heat, and take out the *kombu* just before the water boils.

Katsuobushi is the general name given to dried bonito. Fillet bonito is formed into a hard, dry block in an involved process: the fillet bonito is simmered, then dried for a long time, smoked, and then fermented with fungi for more than 6 months, during which the proteins and nucleotides in the fish are broken down, causing the flavors to intensify. When making "*katsuo-dashi*," a soup stock prepared with *katsuobushi*, Japanese chefs typically shave the dried bonito and then add it to the saucepan as described for *kombu-dashi*.

Japanese *dashi* can be quickly prepared in only 10 min, while it takes a week to prepare high-quality stock or bouillon for French. However, as described above, a long, labor-intensive process is required to produce the *kombu* and *katsuobushi*.

Finally, it should be noted that oils and animal fats are not used in traditional Japanese cuisine, which is very distinct from Western cuisine and is one of the reasons why Japanese cuisine is healthy.

2.2 Discovery of Glutamate as an Essential Component of *Dashi*

2.2.1 Professor Kikunae Ikeda

Food consists of various chemical substances. Chemicals induce different tastes in the mouth and smells in the nose, those senses being collectively called chemical senses. It is not surprising that chemists are interested in the chemical senses. More than 100 years ago, a Japanese scientist, Professor Kikunae Ikeda of the Tokyo Imperial University (now the University of Tokyo), started research on taste. During his stay in Leipzig, Germany, for the study of chemistry from 1899 to 1901, he experienced such foods as asparagus, tomato, cheese, and meat which were not commonly available in those days in Japan. He was so attentive a taster that he found out something common in the complicated taste of those foods, which was quite peculiar and could not be classed under any of the well-defined taste qualities, sweet, salty, sour, and bitter. He also says that the taste is usually so faint and overshadowed by other stronger tastes that it is often difficult to recognize it unless the attention is specially directed towards it.

He was born and raised in Kyoto, one of the most historically important cities in Japan and was the capital of Japan for about a thousand years since 794. The long history of being the center of Japan contributed to creating the typical Japanese cuisine with which he was accustomed. After coming back to Japan, he noticed the existence of the same unique taste in *kombu-dashi*, which he thought to be quite distinctive and not to be classified under any of the conventional four basic taste qualities. He started searching *kombu* for its essential taste substance, and in 1908 he finally isolated a substance as a crystal form whose taste was exactly the same as what he had been detecting. It was L-glutamic acid, an amino acid, which was later proved to exist more abundantly in dried *kombu* than any other foods, and he temporarily named the unique taste "umami" (Ikeda 1909). He mentioned that he used this name because this unique taste was mostly associated with the impression of "*umai*," a Japanese adjective, which means delicious, nice, palatable, or savory. In a later International Meeting held in 1912, New York, however, he proposed "glutamic taste" for this taste quality instead of umami. Although quite accurate and straightforward, this chemical-sounding word, glutamic taste, may be less impactful and hardly acceptable in comparison with umami and might not have been used worldwidely among chefs and gastronomes.

2.2.2 *Glutamic Acid vs. Glutamate*

Ritthausen (1866), a German scientist, had already discovered glutamic acid back in 1866 by acid hydrolysis of wheat gluten, and as cited in Ikeda's paper (Ikeda 1909; Lindemann et al. 2002), Fischer (1906) noted that glutamic acid had a peculiar "bad" taste with weak sourness. If you put a small amount of glutamic acid into your mouth, you will immediately taste sourness because of its dissociated hydrogen ions and gradually you will recognize the unique taste following the buffering action of saliva. Ikeda mixed a small amount of sodium bicarbonate with glutamic acid to neutralize and eliminate sourness, and he served the mixture to people to demonstrate the taste of umami, which resulted in bad reputation because of CO_2 gas formation when dissolved in water. Ikeda soon found that the salt form of glutamate, not glutamic acid itself, elicits the unique taste, umami, without the sour taste because of its neutral pH.

Ikeda showed that various salts of glutamate, e.g., sodium, potassium, magnesium, or calcium, elicit umami with a subtle taste of cation depending on the kind of salt. He confirmed that glutamate, no matter which cations are accompanied, elicits umami. Glutamate actually is present not only in *kombu* but more or less in various foods. Ikeda tried to develop a new flavoring substance on the basis of his finding compatible with the natural taste of foods. He concluded "monosodium L-glutamate" (MSG) to be the best flavoring because it is readily soluble in water and has a pure and strong umami taste, high stability, and absorbs no humidity. MSG is now produced by fermentation technology.

2.2.3 *Global Recognition of Umami*

Ikeda's finding was not known for decades in the West because his original paper was written in Japanese and was not translated into English. Western and American people have been familiar with glutamate-rich ingredients, such as parmesan cheese, tomato, meat, fishes, and chicken soup, and no one had ever noticed a unique taste in them mainly because the taste is so faint and overshadowed by other stronger tastes. Moreover, they have considered that MSG is a "flavor enhancer." Ikeda's 1912 speech in New York reportedly inspired more nap-taking than note-taking (Kasabian and Kasabian 2005). Nevertheless, we owe Ikeda's distinguished and careful talent to detect a unique taste in *kombu-dashi* as well as in cheese, asparagus, tomato, and meat for the discovery of umami.

The predominant free amino acids contained in *kombu-dashi* are glutamate and aspartate, and these two contribute to its umami taste. Both glutamate and aspartate impart the umami taste, but the umami taste intensity of aspartate is only one-tenth of that of glutamate. This means that the major contributor to the umami taste in *kombu-dashi* is glutamate, making *dashi* a very simple or pure umami solution. In comparison, French bouillon and Chinese *tang* contain a wider variety of amino acids that come from their meat and vegetable ingredients, and it is the harmony of these complex components that characterizes the taste of the bouillon and tang. From childhood, Japanese experience umami through dashi, whereas Westerners and the Chinese are used to a more complex taste. This may be another reason why it has taken so long for people in the West and China to recognize the taste, umami.

In 1985, the first international symposium on umami was held in Hawaii where a variety of scientific papers in different fields were presented to suggest and demonstrate that umami is the fifth basic taste along with sweet, sour, bitter, and salty (Kawamura and Kare 1987). There was also a hot discussion concerning how to call the taste of glutamate, and finally they reached a consensus to use umami because of the lack of proper English word synonymous for it. It was not until this symposium that the Japanese word umami has been accepted into the international vocabulary of taste. Umami is more worldwidely known after the registration of "*washoku*" (traditional Japanese cuisine) as an intangible cultural heritage by the United Nations Educational, Scientific, and Cultural Organization (UNESCO) on December 4, 2013, not only in the culinary world but among ordinary people.

2.3 Discovery of Other Umami Substances

After the discovery of MSG, other umami substances were found. In 1913, Ikeda's protégé Shintaro Kodama (1913) found that inosinate (5′-inosine monophosphate, IMP, one of 5′-ribonucleotides) was the key element for deliciousness of *katsuobushi*, which also elicited umami. Subsequent studies have shown that a relatively large quantity is found in animal substances, particularly dried sardines, bonito flakes, and meat. IMP comes from ATP (adenosine triphosphate) through decomposition. Production of IMP proceeds slowly, and the concentration reaches maximum about 10 h after animal's death.

In 1960, Akira Kuninaka (1960) found that guanylate (5′-guanylate monophosphate, GMP, one of 5′-ribonucleotides) had also umami, and this substance was later found to exist abundantly in mushrooms, especially in dried form of *shiitake* mushrooms. Salts of L-glutamate, 5′-inosinate, and 5′-guanylate are collectively called umami substances.

Foods that contain such umami substances abundantly are often called "umami-rich foods." It is noted here that most umami-rich foods are not necessarily eliciting obvious umami taste because umami taste is too faint to be overshadowed by other stronger tastes. Umami-rich natural foods are collectively shown in a review paper (Yamaguchi and Ninomiya 2000). Shortly, umami-rich foods containing glutamate are *kombu*, parmesan cheese, green tea leaves, tomatoes, soy sauce, and *miso* (fermented soybean paste). Inosinate-rich foods are dried bonito, sardines, pork, and chicken. Guanylate-rich foods are dried *shiitake* mushrooms and dried morel mushrooms. The levels of umami increase significantly during the maturation process of food as can be seen in cheese and cured ham. Free glutamate levels increase significantly at varying stages of tomato ripeness. Fully ripened red tomatoes contain a large amount of glutamate. When you wish to make authentic Japanese cuisine outside Japan, it is not easy to use *kombu*, *katsuobushi*, and dried *shiitake* mushrooms. In this case, however, alternative foodstuffs, such as dried tomato instead of *kombu*, dried morel mushrooms instead of dried *shiitake* mushrooms, and chicken breast instead of *katsuobushi*, successfully replace the Japanese *dashi*.

2.4 Synergism

Kuninaka found that there was a taste synergism between glutamate and 5′-ribonucleotide (inosinate or guanylate) (Kuninaka 1960; Kuninaka 1964), i.e., when both types of umami substances are mixed, the intensity of the umami taste is markedly enhanced. Yamaguchi (Yamaguchi 1967), on the basis of relationship between the subjective taste intensity and the mixing ratio of MSG and IMP, found that the maximum umami intensity of their mixture increased up to 7–8 times larger than the umami intensity of each component. Yamaguchi and Kimizuka (Yamaguchi and Kimizuka 1979) also demonstrated in terms of the threshold concentration that the detection threshold for MSG (0.012%, wt/v) is low enough to be used as a seasoning, but is larger than that for tartaric acid (0.00094) or quinine sulfate (0.000049). However, the threshold of MSG is markedly lowered (0.00019) in the presence of IMP (in 0.25% IMP solution). This is due to a synergistic taste effect between MSG and IMP. In another expression of synergism, Yamaguchi (Yamaguchi 1967; Yamaguchi et al. 1971) reported that the synergistic effect between them could be expressed quantitatively by equation: if u g/100 ml of MSG is mixed with v g/100 ml of IMP (or GMP) solution, the taste intensity of the resulting solution in terms of the concentration of MSG alone, y g/100 ml, can be expressed as $y = u + 1218uv$ for IMP and $y = u + 2800uv$ for GMP.

Long before Kuninaka's scientific discovery about the synergy of umami by mixing glutamate and 5′-ribonucleotides such as inosinate or guanylate, Japanese people had noticed a technique of making a *dashi* with enhanced deliciousness, that is, a *dashi* called "*ichiban-dashi*" (or combined *dashi*) made from two umami-rich ingredients, *kombu*, which is rich in glutamate, and dried bonito, which contains a large quantity of inosinate, to increase umami up to about eight times more than when with glutamate alone.

2.5 Dual Function of Glutamate

As already described, Ikeda noticed that glutamate had a unique taste that does not fall into the conventional four basic taste qualities and named the taste umami. This term was coined by him on the basis of recognition that glutamate imparted a familiar and highly desirable flavor to foods. A confusion or misunderstanding that "umami is delicious" comes from such a background of naming, i.e., it must be stressed here that umami denotes a unique taste that alone is not necessarily delicious (Yamaguchi and Takahashi 1984; Beauchamp and Pearson 1991). As Halpern (Halpern 2002) suggested, it is appropriate for glutamate to speak of "glutamic taste" rather than "umami" in a similar fashion that salt (NaCl) is salty (or has salt taste or saltiness). However, as shown in the following sections, when mixed with other ingredients in cooking, glutamate makes cuisine delicious; otherwise it is not so appetitive (Yamaguchi 1998) (Fig. 2.1).

Generally speaking, there are two distinctive functions of a single tastant, i.e., one is to elicit a particular taste and the other is to influence the flavor of tastants in a food. It is not easy to explain scientifically why glutamate alone is not delicious but it makes food delicious. Recently, Yasumatsu et al. (2015) suggest that T1R1/T1R3-expressing cells mainly contribute to umami synergism and partly to sweet sensitivity, while mGluRs are involved in the detection of umami compounds, suggesting that the former cells are concerned with palatability-inducing action and the latter cells are related to elicitation of unique taste of umami. This idea is interesting in the sense that distinction of quality and hedonics are already made at the peripheral receptor level. In summary, glutamate has a unique taste called umami, which

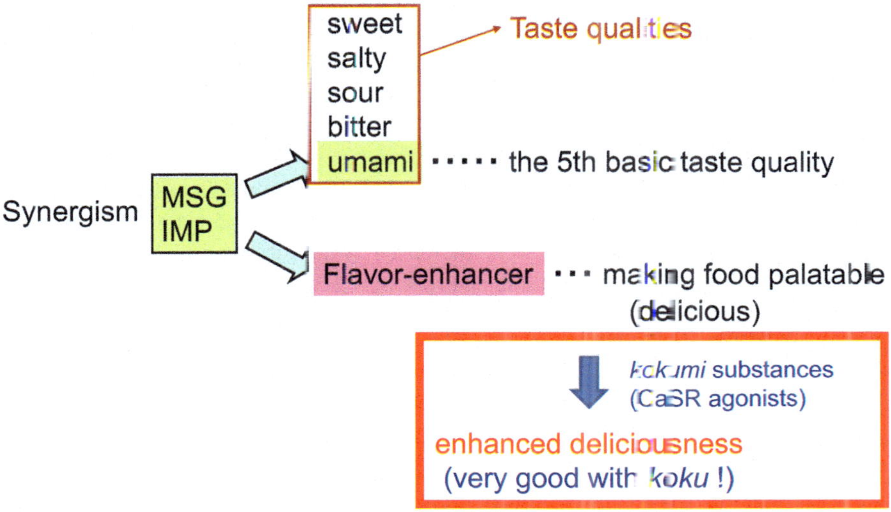

Fig. 2.1 Functions of MSG. MSG has two functions: one is to induce umami and the other is to make food delicious. MSG shows synergism with IMP. Further addition of *kokumi*-inducing substances, agonists for CaSR (calcium-sensing receptor), induces enhanced deliciousness

is recognized in the primary gustatory area of the cortex, but when mixed with ingredients of food, it elicits a flavor-enhancing effect to make food delicious and pleasant through processing in the affection-related brain areas.

2.6 Flavor-Enhancing Action of Glutamate

Why does food become delicious when glutamate is present? Is it due to the unique taste (umami) of glutamate or due to a flavor-enhancing effect (or seasoning effect) of glutamate indifferent from umami taste? Although the latter may be correct as described in the previous section, it is not easy to decide which is correct because it is impossible to separate umami taste from its seasoning effect because these two usually occur concomitantly. A clue to understanding umami taste can be found in the notion as already described that umami itself is not necessarily a good taste like sugar (sweet taste) but is similar to the case of salt, i.e., salty taste is not necessarily good, but a certain concentration of salt is essential in making food delicious. Notably, you can enhance deliciousness with the addition of near-threshold or low concentrations of MSG (Yamaguchi 1998) which may not elicit apparent umami taste. However, if you add an excess amount of MSG, the food would be avoided because of the heaviness of too strong umami taste and excess salivation, leading to diminution of palatability.

In accordance with the above statement, Yamaguchi and Takahashi (1984) showed that umami, the taste of MSG, is neutral or not palatable by itself in terms of pleasantness, but it makes a variety of foods palatable when properly used. However, simple combinations of MSG with other basic tastes remain neutral or give negative scores, meaning that glutamate does not influence the pleasantness in simple binary mixtures of basic tastes. On the other hand, when a beef consommé was used in her experiment as a food, the overall taste intensity of the food was properly increased by the addition of MSG at the concentration of as low as 0.2% which is around the recognition threshold (Yamaguchi 1998), suggesting that interaction of MSG with unknown substances in beef consommé is essential in flavor enhancement. As shown in Fig. 2.2, concerning each of the classical four basic tastes, sweetness and saltiness were increased and sourness and bitterness were slightly decreased. The most notable changes were major increases of flavor characteristics, which may partly be expressed by “amplitude” in Western cultures, i.e., “thickness,” “continuity,” “mouthfullness,” “impact,” and “mildness.” These flavor-enhancing characteristics are collectively called “*koku*” in Japanese. *Koku* will be explained precisely in the following sections. Thus, glutamate expands the overall deliciousness of food.

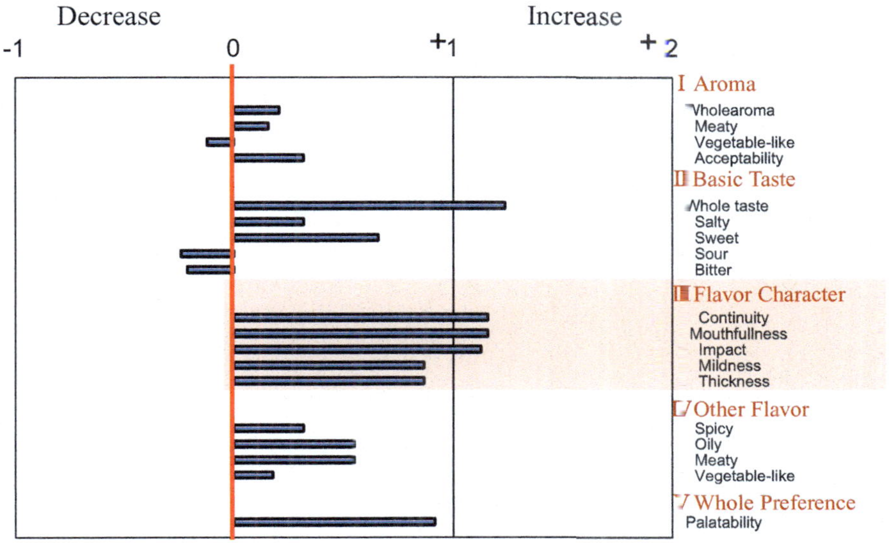

Fig. 2.2 Addition of MSG on flavor profile of beef consommé (Modified from Yamaguchi 1998)

2.7 *Koku*

Deliciousness has an important role to play in enhancing consumption of food. It is natural to seek stronger deliciousness in cuisine. Partly because the flavor of Japanese *dashi* is subtle and mild, Japanese people had noticed a technique of making *dashi* more delicious, that is, a *dashi* called "*ichiban-dashi*" as described above. Some Japanese people may say that *ichiban-dashi* has *koku* in comparison to simple *kombu-* or *katsuo-dashi*. Further, putting soy sauce or *miso*, both are rich in various amino acids, into the *ichiban-dashi* will boost the recognition of *koku*. *Koku* is a Japanese word literally derived from "strong," "rich," or "concentrated." Generally, Japanese use the word *koku* on a daily basis when they evaluate a stronger deliciousness of food. They often use this word in phrases, such as "this food is very delicious because of the *koku* in it."

An essential condition for inducing *koku* is the existence of glutamate and other complex chemicals including rich amino acids in food. It is said, for example, that Cheddar cheese aged 9 months has more *koku* than the aged 2 months because the former has rich chemical reactions or decomposition products. Likewise, vintage wines have more *koku* than young wines such as Beaujolais Nouveau because the former has rich compositions, and soup cooked for 6 h has more *koku* than soup cooked for 1 h, for the same reason. Thus, *koku* is the most appropriately used when we enjoy the odor, texture, and color as well as the taste of food containing complex compounds generally obtained after maturation, fermentation, aging, curing, drying, and slow cooking. *Koku* for wine is customarily expressed as "body" when people taste wine. Similarly, people may use "richness" instead of *koku* for some foods or drinks.

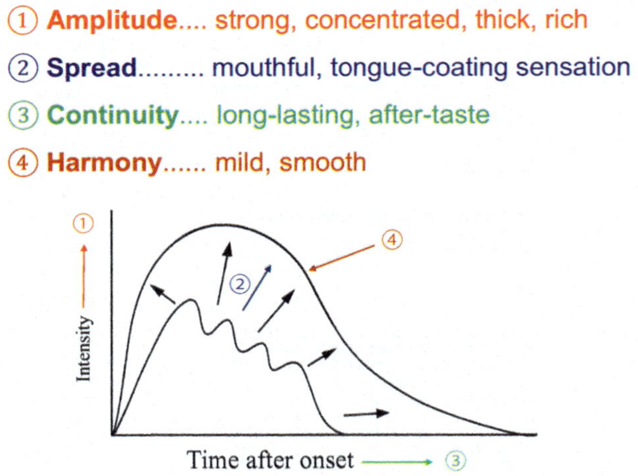

Fig. 2.3 An image of *koku* (or *kokumi*) with four *koku* attributes. Under the influence of *kokumi* substances, food with umami substances becomes very delicious with the *koku* attributes

Koku is a conceptual word used for edibles implying extensive flavor characteristics induced by multiple sensory interaction in many cases leading to a state of strong palatability. Therefore, *koku* is not one of the sensory modalities, such as taste, smell, touch, pain, vision, and audition, nor a qualitative aspect of sensations, such as sweet, sour, salty, bitter, warm, cold, red, and green. *Koku* is related to the quantitative aspect of sensations and positive hedonics. To evaluate *koku* scientifically in sensory test, attributes of *koku* are used, such as thickness (concentrated, amplitude, strength), continuity (aftertaste), mouthfullness (tongue-coating sensation), mildness (smoothness, balance, harmony), depth (richness, complexity), and punch (impact). Thickness refers to rich complexity, continuity refers to long-lasting sensory effects including an increase of aftertaste, and mouthfullness refers to sensory reinforcement or the spread of sensation throughout the whole mouth. These attributes could be explained by the *koku* image as shown in Fig. 2.3, i.e., thickness is shown by the increase of ordinate; continuity, by the increase of abscissa; mouthfullness, by the expansion of the area; mildness, from uneven curve to smooth curve; depth, from small to large numbers of symbols; and punch, steep increase of the onset of the curve.

2.8 *Koku* vs. *Kokumi*

As described above, combinations of complex sensory modalities and qualities induce *koku*. However, if one or more specific key substances coexist in food containing umami substances, the food becomes explosively delicious with *koku*. Such key substances, which are called "*koku*-inducing substances," include glycogen, fat, oil, alliin, glutathione, sulfur-containing compounds, some specific peptides, Maillard reaction products, amino acids, heated products of gelatin, and tropomyosin. Only a single sensory modality could be effective in inducing *koku*. For example, such texture as increased viscosity or thick consistency of food is an important factor in inducing *koku*. Certain aromas are known to be very effective in eliciting *koku*. For example, among the characteristic odorants of celery, phthalides, such as sedanenolide, 3-n-butylphthalide, and sedanolide, contribute to flavor enhancement of chicken broth (Kurobayashi et al. 2008). When glutamate is given in combination with a consonant, savory aroma (vegetable), the resulting flavor becomes much more pleasant (Rolls 2009; Nishimura et al. 2016).

Koku can be induced by interaction only among tastes. For this phenomenon, a pioneering study was performed by Ueda et al. in 1990. According to their paper (Ueda et al. 1990), they say that "addition of a small amount of extract of garlic to umami solution (MSG + IMP) enhances its flavor characters, such as continuity, mouth-fullness and thickness." These characters are the same as those described above for *koku*; however, they called these flavor characters as "*kokumi* flavors." This coined term *kokumi* is a compound word of "*koku*" plus '*mi*" ("*mi*" denotes taste in Japanese). Introduction of this term, however, makes the "*koku* world" very confused. Japanese never use *kokumi* in daily life, while scientists in the relevant area are now using *kokumi* rather than *koku* because several researchers followed this naming and essentially no researchers have used *koku*. Even the expression of "*kokumi* taste" is appearing in scientific papers on taste (Ohsu et al. 2010). It is noted here that *kokumi* is neither a kind of unique taste as has been so in the case of umami nor a different kind of *koku*. Again *kokumi* is coined to denote *koku* which is induced by modification of taste information by adding some substances even without eliciting taste. Such substances are called "*kokumi* substances," which are equivalent to *koku*-inducing substances. When you use *kokumi*-inducing substances, however, you should bear in mind that these substances exert their function after binding to *kokumi* receptors described below.

Glutathione, one of the *kokumi* substances, was investigated by Ueda et al. (1997) in humans and by Yamamoto et al. (2009) in mice. Glutathione is a tripeptide with glutamic acid, cysteine, and glycine (L-γ-glutamyl-L-cysteinylglycine, simply γ-Glu-Cys-Gly) that is widely included in foodstuffs such as meat, seafood, and wine. Yamamoto et al. conducted behavioral and electrophysiological experiments to explore mixture effects between glutathione and umami substances in mice. They found that the addition of glutathione to IMP, but not to monopotassium glutamate (MPG), increased preference for the solution (Fig. 2.4). Neural responses of the taste nerves to the mixture of glutathione and IMP, but not glutathione and MPG, showed synergism (Fig. 2.5).

They suggested that glutathione enhanced preference for umami solutions containing 5′-ribonucletides rather than glutamate. There is a possibility of existence of species difference which should be clarified in future studies. In a human sensory test, Ueda et al. (1997) also used the simplest sample, an umami solution containing 0.05% each of MSG and 5′-inosinate and reported that this peptide increased the flavor characteristics of this umami solution. They reported that the increased flavor

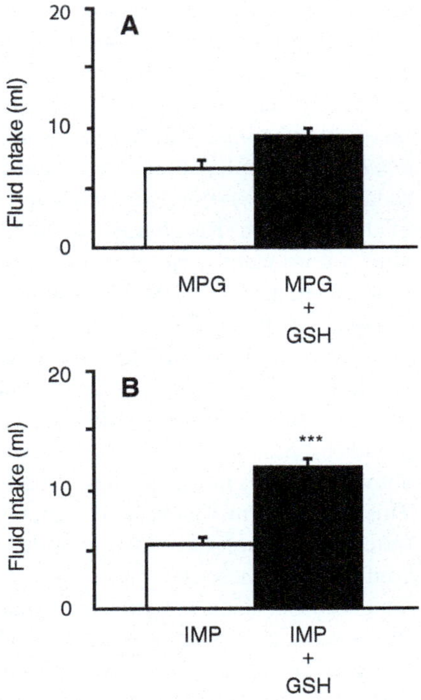

Fig. 2.4 Mean volume of fluid intake ± SE for 0.1 M monopotassium chloride (MPG) with or without 1 mM glutathione (GSH) (**a**) and 1 mM inosine monophosphate (IMP) with or without 1 mM GSH (**b**) in the two-bottle preference test in mice. A pair of liquids in each graph was presented simultaneously for 48 h. Modified from Yamamoto et al. (2009). ***$P < 0.001$ (t-test)

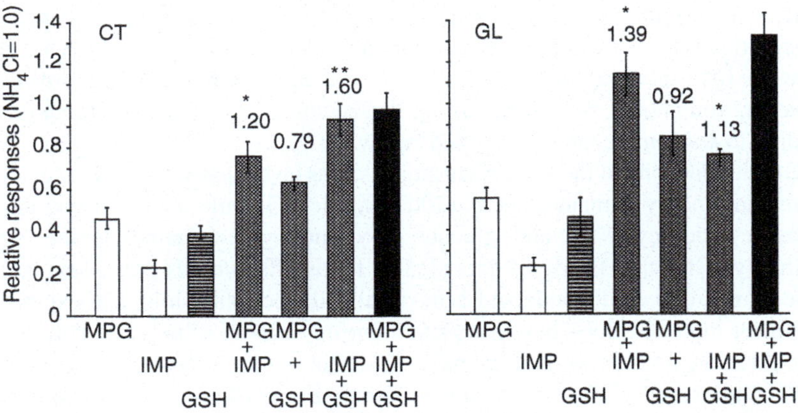

$$\text{Synergism: } 1.0 < \frac{\text{Response to a mixture}}{\text{Sum of responses to individual components}}$$

Fig. 2.5 Mean relative responses ± SE to MPG, IMP, GSH, and their mixtures in the chorda tympani (CT), which innervates the anterior part of the tongue, and in the glossopharyngeal nerve (GL), which innervates the posterior part of the tongue. Values above the bars show potentiation ratios calculated with the equation shown below the graphs. Modified from Yamamoto et al. (2009). *$P < 0.05$, **$P < 0.01$ (t-test between the mixture response and the sum of responses to individual components) (or enhanced deliciousness or savoriness) of the umami solution could be expressed by such terms as continuity, thickness, and mouthfullness. Judging from the result of this experiment, a prototypical meaning of *koku* may simply be interpreted as highly enhanced state of deliciousness originally produced by umami substances.

2.9 CaSR Agonists

SanGabriel et al. (2009) have shown that a receptor called the calcium-sensing receptor (CaSR) exists in type 2 and type 3 taste cells. The CaSR expressed in the parathyroid glands and kidneys is well known to play a central role in extracellular calcium homeostasis, and CaSR expressed in other tissues also play some biological functions. CaSR agonists are Ca ions, basic peptides such as protamine and polylysine, and aromatic amino acids such as histidine, tryptophan, and phenylalanine. Ohsu et al. (2010) found that γ-glutamyl-valyl-glycine (γ-Glu-Val-Gly) was the most active CaSR agonist among 46 γ-glutamyl peptides including glutathione and induced the most prominent *koku* when 0.01% γ-Glu-Val-Gly was mixed independently with 3.3% sucrose, 0.9% NaCl, or 0.5% MSG. These *kokumi* substances bind CaSR, to increase sweet, salty, and umami responses and to induce the *koku* effects in a human sensory test (Ohsu et al. 2010; Maruyama et al. 2012). CaSR is now established as a receptor of *kokumi* substances. Interestingly, even the subthreshold concentrations of these substances are effective, indicating that the taste qualities of these substances are not required (Ohsu et al. 2010). γ-Glu-Val-Gly is not an artificial product but is contained in natural foods, such as commercial fish sauces (Kuroda et al. 2012) and commercial soy sauces (Kuroda et al. 2013).

Our recent study suggests that GPRC6A is another candidate receptor of *kokumi* substances. Ornithine is the potent agonist for this receptor (Mizuta and Yamamoto 2018, unpublished).

2.10 Interaction Between Glutamate and *Kokumi* Substance

Thus, recent studies in the field of food chemistry, molecular biology of taste receptors, and psychophysics are revealing the nature of palatability of food in terms of *koku* as well as umami. Hypothetically, generalists in a subset of taste cells sending appetitive information are strongly activated by *kokumi* substances through CaSR receptors. It has been a common phrase that umami substances induce deliciousness when added to ingredients during cooking. But the reverse expression may be more appropriate, i.e., umami responses are explosively increased under the presence of certain *kokumi* substance(s) existing in ingredients.

The concept of *koku* might be understood by most Japanese who eat cuisine with mild flavors created by *kombu*- and/or *katsuo-dashi*. Soup stock in China and Western countries has complex ingredients including glutamate, *koku*-inducing substances, and *kokumi* substances and elicits enhanced flavor, i.e., *koku* already exists in the soup stock. Therefore, there is a possibility that those people who are already accustomed to enhanced umami stock may be hard to understand what *koku* means. I would like to propose to those who are unfamiliar with *koku* to think of the taste of milk or actually to taste milk (preferably human breast milk). It contains enough amounts of glutamate, fat, lactose, and calcium, and interactions among these substances induce not only simple deliciousness but complex, deep, and enriched sensation with satisfaction, which may be similar to what Japanese imagine about *koku*.

2.11 Summary

1. Monosodium glutamate (MSG) has a dual function: one is to elicit a unique taste called umami and the other is to enhance deliciousness of food.
2. Deliciousness of cuisine basically comes from seasoning effects of MSG.
3. *Kombu-dashi* is rich in MSG and is the simplest flavor enhancer commonly used in daily life in Japan.
4. *Ichiban-dashi* induces stronger deliciousness because of synergy between *kombu-dashi* (glutamate) and *katsuo-dashi* (inosinate).
5. Addition of *miso* or soy sauce, which is rich in various amino acids, to *ichiban-dashi* induces much stronger deliciousness with thickness, mouthfullness, and continuity, which are often referred to as *koku* among Japanese.
6. *Koku* could be simply defined as a state of enhanced deliciousness.
7. Western and Chinese soups already have *koku* in themselves.
8. *Koku*-inducing substances, including various substances with different modalities, enhance deliciousness of food with the *koku* attributes.
9. *Kokumi* is a coined word to denote *koku* induced by certain chemical substances, such as glutathione and Ý-Glu-Val-Gly, which are agonists for CaSR.
10. These substances are called *Kokumi* substances, which enhance deliciousness of food through taste interactions with umami substances.

Acknowledgments The Umami Information Center (UIC) in Tokyo provides two booklets, "Umami" and "Umami the World," which are very informative and helpful in preparation for the present manuscript. This review paper was supported by JSPS KAKENHI (Grant Number 17K00835) and a grant for Project Research from Kio University.

References

Beauchamp GK, Pearson P (1991) Human development and umami taste. Physiol Behav 49:1009–1012

Halpern BP (2002) What's in a name? Are MSG and umami the same? Chem Senses 27:845–846

Ikeda K (1909) On a new seasoning. J Tokyo Chem Soc 30:820–836. (in Japanese)

Kasabian D, Kasabian A (2005) The fifth taste. Columbia University Press, New York, p 16

Kawamura Y, Kare MR (eds) (1987) Umami: a basic taste. Marcel Dekker, New York

Kodama S (1913) On a procedure for separating inosinic acid. J Tokyo Chem Soc 34:751–757. (in Japanese)
Kuninaka A (1960) Studies on taste of ribonucleic acid derivatives. J Agric Chem Soc Jpn 34:487–492. (in Japanese)
Kuninaka A (1964) The nucleotides, a rationale of research on flavor potentiation. Symposium on flavor potentiation. AD Little, Cambridge, MA, pp 4–9
Kurobayashi Y et al (2008) Flavor enhancement of chicken broth from boiled celery constituents. J Agric Food Chem 56:512–516
Kuroda M et al (2012) Determination and quantification of γ-Glutamyl-valyl-glycine in commercial fish sauces. J Agric Food Chem 60:7291–7296
Kuroda M et al (2013) Determination and quantification of the *kokumi* peptide, γ-glutamyl-valyl-glycine, in commercial soy sauces. Food Chem 141:823–828
Lindemann B, Ogiwara Y, Ninomiya K (2002) The discovery of umami. Chem Senses 27:847–849. (English translation)
Maruyama Y, Yasuda R, Kuroda M, Eto Y (2012) *Kokumi* substances, enhancers of basic tastes, induce responses in calcium-sensing receptor expressing taste cells. PLoS One 7:e34489
Nishimura T et al (2016) Umami compounds enhance the intensity of retronasal sensation of aromas from model chicken soups. Food Chem 196:577–583
Ohsu T et al (2010) Involvement of the calcium-sensing receptor in human taste perception. J Biol Chem 285:1016–1022
Rolls ET (2009) Functional neuroimaging of umami taste: what makes umami pleasant? Am J Clin Nutr 90:804S–813S
San Gabriel A, Uneyama H, Maekawa T, Torii K (2009) The calcium-sensing receptor in taste tissue. Biochem Biophys Res Commun 378:414–418
Ueda Y, Sakaguchi M, Hirayama K, Miyajima R, Kimizuka A (1990) Characteristic flavor constituents in water extract of garlic. Agric Boil Chem 54:163–169
Ueda Y, Yonemitsu M, Tsubuku T, Sakaguchi M, Miyajima R (1997) Flavor characteristics of glutathione in raw and cooked foodstuffs. Biosci Biotechnol Biochem 61:1977–1980
Yamaguchi S (1967) The synergistic taste effect of monosodium glutamate and disodium 5′-inosinate. J Food Sci 32:473–478
Yamaguchi S (1998) Basic properties of umami and its effects on food flavor. Food Rev Int 14:139–176
Yamaguchi S, Kimizuka A (1979) Psychometric studies on the taste of monosodium glutamate. In: Filer LJ et al (eds) Glutamic acid: advances in biochemistry and physiology. Raven Press, New York, pp 35–54
Yamaguchi S, Ninomiya K (2000) Umami and food palatability. J Nutr 130:921S–926S
Yamaguchi S, Takahashi C (1984) Hedonic function of monosodium glutamate and four basic taste substances used at various concentration levels in simple and complex systems. Agric Biol Chem 48:1077–1081
Yamaguchi S, Yoshikawa S, Ikeda S, Ninomiya T (1971) Measurement of the relative taste intensity of some L-α-amino acids and 5′-njucleotides. J Food Sci 36:846–849
Yamamoto T, Watanabe U, Fujimoto M, Sako N (2009) Taste preference and nerve response to 5′-inosine monophosphate are enhanced by glutathione in mice. Chem Senses 34:809–818
Yasumatsu K et al (2015) Involvement of multiple taste receptors in umami taste: analysis of gustatory nerve responses in metabotropic glutamate receptor 4 knockout mice. J Physiol 395:1021–1034

Chapter 3
The Quest for Umami

Ole G. Mouritsen

Abstract More than a century after Kikunae Ikeda in 1909 proposed a fifth taste quality, *umami*, as a basic taste, many people and even chefs, in particular outside Japan, still struggle with a clear perception of umami in their food. Often umami is used incorrectly as synonymous with deliciousness and a lucky combination of the four classical basic tastes along with a pleasant mouthfeel. Umami is now known to be elicited by the stimulation of umami receptors in the taste buds by the interaction with free amino acids, in particular glutamate, often in synergy with free 5′-ribonucleotides, such as inosinate, adenylate, and guanylate. Umami is an important component in the *koku* attribute by enhancing continuity and mouthfullness of the overall taste experience. In order to determine if certain foodstuff elicit true umami, we have undertaken a quantitative analysis of the free amino acid contents of various foodstuff and food preparations often claimed to have umami, in particular fermented sauces, seaweeds, cephalopods, and sous vide-prepared meat. We find that in some cases these kinds of foodstuff, even if delicious, do actually contain very little free glutamate.

Keywords Umami · *Kokumi* substance · Glutamate · Gastronomy · Fermentation · Fish · Seaweed · Dashi · Squid · Sous vide

3.1 Introduction

Although being recognized widely, even throughout the Western world, many people have difficulty coming to grasp with the term 'umami' as a basic taste sensation. One reason is, as already pointed out in the pioneering work by Ikeda (2002 [1909]), that this quality often is hidden under the other basic tastes, and that it has a time course in the mouth which often overlaps with mouthfeel, mouthfullness, *kokumi*, complexity, etc. Another reason is that most world cuisines, apart from the classical

O. G. Mouritsen (✉)
Department of Food Science, Taste for Life, Design and Consumer Behaviour,
University of Copenhagen, Frederiksberg C, Denmark
e-mail: ole.mouritsen@food.ku.dk

T. Nishimura, M. Kuroda (eds.), *Koku in Food Science and Physiology*,
https://doi.org/10.1007/978-981-13-8453-0_3

Japanese cuisine, do not have a single ingredient in the kitchen, which is very pure in umami except for MSG (monosodium glutamate). In contrast, the Japanese kitchen has the unique soup broth *dashi* as well as an understanding and general appreciation of the principle behind it (Antony et al. 2014; Mouritsen and Styrbæk 2014). Even if Western foodstuff like matured and fermented cheeses, cured ham, sardine paste, ketchup, etc., are rich in umami compounds, their taste is seldom by the ordinary eater characterized by the term umami, possibly also because umami as an expression is not yet firmly encoded in most non-Japanese languages (O'Mahony and Ishii 1986).

Interestingly, the antique Mediterranean cuisine was to a large extent based on the use of a condiment, *garum* (Curtis 2009; Grainger 2010), that is basically a fermented fish sauce (Mouritsen et al. 2017) which is both an umami and salt provider, in much the same way as Asian fish sauces, such as the classical Japanese and Vietnamese fish sauces *ishiri* and *nuoc mam tom cha*. Similarly, many coastal regions in Europe and the Americas have traditionally used various kinds of seaweeds as food and flavouring agents (Mouritsen 2013; Mouritsen et al. 2019a; Pérez et al. 2018), although not to the same extent and in the same way as Japanese for centuries have used the brown seaweed *konbu* (*Saccharina japonica*, *S. longissima*, as well as several other of the *Saccharina* genus) as the motherlode of umami in *dashi*. Moreover, just like high-quality sake is described by umami via its contents of free amino acids, derived from the autolyzed yeast and the lees (*sake-kasu*), Western champagnes and other sparkling wines are often by connoisseurs associated with umami due to the production methods involving the wine maturing along with the yeast sediment. As a final example, it has often been claimed that slow cooking of meat, e.g., by simmering and sous vide techniques, leads to more umami.

Despite these examples of Western foodstuff possibly associated with true umami sensation via their chemical composition, it remains in most cases to be demonstrated quantitatively that these types of foodstuff actually contain the specific compounds that bind to the umami receptor and signal true umami sensation to the taste centre in the brain. And even if this is the case, Western languages do not traditionally use a specific term like umami to describe the common taste of such products like fish sauce, cheese, tomatoes, and cured ham. Beyond this, a further complexity arises due to possible interactions between aroma compounds and the perceived intensity of the umami taste (Mouritsen et al. 2019b).

In the present paper, we review some results of recent work inspired by the question: is the claim of umami in particular foodstuffs actually supported by the chemical composition of the taste compounds; specifically does the foodstuff in question in fact contain appreciable amounts of free glutamate? Most of the work mentioned is concerned with food from the ocean: fish, seaweeds, and cephalopods. Marine food sources are among the very best candidates for umami-rich food (Komata 1990; Mouritsen 2016, 2017) with high levels of free amino acids and nucleotides in the raw materials (Maga 1983; Ninomiya 1998).

3.2 Umami of Fermented Sauces

The most powerful techniques to produce umami compounds involve fermentation by enzymes, moulds, bacteria, and other microorganisms, which break proteins down into peptides and free amino acids and nucleic acids into free nucleotides. Fermented products are traditionally an important component of most food cultures (Katz 2012), such as fermented milk (cheeses, sour milk products), fermented beans (soy sauce, miso, *nattō*, *douchi*, *furu*), fermented grapes, fruit, berries, and grains (wine, beer, cider, brandy, bread), fermented vegetables (*sauerkraut*, *kimchi*, *tsukemono*), fermented leaves (tea), fermented fish (*katsuobushi*, fish- and shellfish sauces, Swedish *surströmming*), and fermented meats (sausages cured hams). Only in a few cases do the process lead to both types of umami compounds in appreciable amounts, and in most cases the end product can be rich in glutamate but contain very little free nucleotides.

Fermentation of fish is a particularly interesting example where quite often products like fish sauces are high in glutamate, but whatever free inosinate may be present in the fresh fish, such as mackerel, anchovies, and sardines, or formed during the fermentation is broken down in the dead fish with the result of formation of inosine or hypoxanthine (Gill 1990; Howgate 2006) unless special preparation techniques are invoked. This is the case for the production of *katsuobushi* (skipjack tuna) and *niboshi* (small sardines) where measures have been taken at and immediately after the catch as well as in the subsequent production process in order to conserve inosinate. Alternatively the fish could be cooked, dried, and marinated immediately after catching (Maga 1983), which is the case in the production of Italian anchovy paste. In both cases, the proteolytic enzymes that could produce glutamate during subsequent fermentation together with the enzymes that can break down inosinate are denatured, leading to a product with little free glutamate but optimal inosinate content.

Therefore, fermented fish sauces owe their umami taste to free glutamate. The same is normally the case with other fermented protein-rich foodstuff. In a recent study we performed a comparative analysis of the free glutamate content in a series of fermented sauces, including not only experimental (cf. Fig. 3.1) and commercial fish sauces but also sauces of fermented squid, insects, game, and pulses (Mouritsen et al. 2017). Some of the resulting data are given in Table 3.1. In none of these samples did we find any appreciable amounts of free nucleotides. It is noted from the table that the contents of glutamate vary dramatically. One of the experimental *garums*, Quick-and-Easy Garum, is not a true *garum* in the sense that it is not fermented but only a cooked extract of whole mackerel. It is produced by an ancient recipe for a sort of quick-and-easy *garum* for the benefit of 'the busy Roman housewife' (Dalby 2011). The recipe involves cooking fish and innards with salt in water, in this case from fresh mackerel, for about 2 h. Hence, its glutamate content is low and similar to that in the fresh fish.

Fig. 3.1 Left: Preparation of *garum* from whole fresh mackerel in salt. Right: Filtered *garum* made from mackerel fermented for 3 months at 28 °C (permission to print by courtesy of Jonas Drotner Mouritsen)

Table 3.1 Content of free glutamate (mg/100 g) in some selected commercial and experimental fermented sauces (Mouritsen et al. 2017)

Fermented sauce	Glutamate	Origin
Delfino (sardines)	3792	Delfino Colatura di Alici di Cetara, Italy
Flor-de-Garum (anchovies)	2161	Flor-de-Garum, Cádiz, Spain
Red Boat 40 N (anchovies)	650	An N Cuong Co, Vietnam
Experimental *garum* (sprats)	612	Jacob Sørensen, Denmark
Ika no Ishiru (squid guts)	581	Shinbo-minoru, Ishikawa, Japan
Experimental *garum* (grasshopper)	304	Nordic Food Lab, Denmark
Experimental *garum* (yellow peas)	218	Nordic Food Lab, Denmark
Experimental *garum* (pheasant)	183	Nordic Food Lab, Denmark
Quick-and-Easy Garum (mackerel)	15	STYRBÆKS, Denmark

The fish and squid *garums* have the highest amounts of free glutamate, presumably facilitated by the high enzymatic activity of their own intestinal enzymes. The experimental garums of grasshopper, peas, and pheasant are all fermented by aid of a barley *koji* mother (Mouritsen et al. 2017). The resulting fermented sauces are somewhat lower in glutamate than those from seafood.

The fermented sauces were also characterized by quantitative sensory evaluation using a trained panel of tasters in order to reveal if tasters can detect high levels of free glutamate as an umami sensation (Mouritsen et al. 2017). The sensory analysis found a fairly good prediction of sensory properties from the chemical characterization of the sauces. However, the relationship between the actual free glutamate concentration and intensity of umami taste is not simple. Therefore, in such chemically complex solutions like these fermented sauces, there may be other sensory cues from, e.g., aroma compounds that interfere with the perception of the umami-tasting compounds.

3.3 Umami from Seaweeds: *Dashi*

Umami was originally proposed by Kikunae Ikeda to describe the unique taste of Japanese *dashi*, and in the original work form 1909 he described that the taste is caused by free glutamate which he discovered was present in large amounts in the brown seaweed *konbu* (*Saccharina japonica*) (Ikeda 2002 [1909]). Since then, monosodium glutamate (MSG), also known as the third spice, has played a similar role for umami as sodium chloride has for salty. In a traditional *dashi*, the pure *konbu* extract (*konbu-dashi*) is supplemented with an extract of *katsuobushi* or dry shiitake that contribute respectively free inosinate and free guanylate, thereby providing for full umami synergy (Zhang et al. 2008). The closest Western or Chinese analogue of Japanese *dashi* is common soup stock prepared from vegetables (glutamate) and meat (inosinate) (Mouritsen and Styrbæk 2014). However, it would be natural to ask the question if other seaweeds than *konbu* could also be used as a component for a *dashi*.

Inspired by the trend in the so-called New Nordic Cuisine, we undertook a study of the umami potential of various seaweeds sourced from Nordic waters (Mouritsen et al. 2012) and discovered that the red seaweed dulse (*Palmaria palmata*), cf. Fig. 3.2 (Mouritsen et al. 2013), was rich in free glutamate and demonstrated that extracts of dulse provide for a very tasty and umami-rich *dashi* that could be used not only as a soup stock but also to enrich the flavour of other dishes, like fresh cheeses, bread, and ice crème (Mouritsen et al. 2012). We have recently extended this study to a major comparative physicochemical analysis of the free amino contents of 20 different species of brown seaweeds used for human consumption from around the word, including the 12 genera *Nereocystis*, *Macrocystis*, *Laminaria*, *Saccharina*, *Undaria*, *Alaria*, *Postelsia*, *Himanthalia*, *Ecklonia* (formerly *Eisenia*), *Sargassum*, *Fucus*, and *Codia* (Mouritsen et al. 2019b).

Table 3.2 shows a selection of results for free glutamate contents for some brown (Mouritsen et al. 2019b) and red seaweed species (Mouritsen et al. 2012, 2013). It is seen that the contents vary dramatically for the different species. In addition to

Fig. 3.2 Dried fronds of the red seaweed dulse (*Palmaria palmata*) that is rich in free glutamate and suitable for preparing *dashi* (Permission to print by courtesy of Jonas Drotner Mouritsen)

Table 3.2 Content of free glutamate (mg/100 g) in some selected aqueous seaweed extracts (*dashi*)

Seaweed extract (*dashi*)	Glutamate	Origin
Ma-konbu (*Saccharina japonica*)	37[a]	Hokkaido, Japan
Hidaka-konbu (*Saccharina japonica*)	18[a]	Hokkaido, Japan
Dulse (*Palmaria palmata*)	10[b]	Stykkisholmur, Iceland
Graceful red weed (*Gracilaria verrucosa*)	6[b]	Grenå, Denmark
Bull kelp (*Nereocystis luetkeana*)	3[a]	Vancouver Island, Canada
Oarweed (*Laminaria digitata*)	1[a]	Grindavik, Iceland
Winged kelp (*Alaria esculenta*)	1[a]	Grindavik, Iceland
Sugar kelp (*Saccharina latissima*)	1[a]	Lillebælt, Denmark
Bladderwrack (*Fucus vesiculosus*)	1[a]	Lillebælt, Denmark
Wakame (*Undaria pinnatifida*)	0[a]	Shimane, Japan

All data are based on extraction at 60 °C of 5 g dry seaweed in 250 g water
[a]Mouritsen et al. (2019b)
[b]Mouritsen et al. (2012)

this variation there is also a substantial variation for the individual species depending on the place and time of harvest.

Without comparison, the different Japanese variants of *konbu* have by far the highest contents of free glutamate. Other local variants of *Saccharina japonica* not listed, such as *Rishiri-konbu* and *Raushu-konbu*, are in between *Ma-konbu* and *Hidaka-konbu*, but there are some differences for different harvesting sites and vendors (Japanese Culinary Academy 2016), a fact that influences the market price.

Table 3.2 shows that the glutamate content in a *dashi* prepared from the red seaweed dulse (*Palmaria palmata*) using the same extraction techniques as for *konbu* is comparable to that in some of the lower quality of *konbu-dashi*. In fact we have found in an earlier study (Mouritsen et al. 2012) that a *dashi* using farmed Danish dulse could contain up to 40 mg/100 g. Noting that another red seaweed graceful red weed (*Gracilaria verrucosa*) has more free glutamate than the rest of the brown species investigated in Table 3.2 suggests that it may be worthwhile to study further the umami potential of other red species. It should also be pointed out that red seaweeds like dulse have very low content of iodine compared to most brown seaweed species, in particular members of the Laminariales order (Teas et al. 2004; Mouritsen et al. 2019b). Hence from the point of view of reducing the human exposure to iodine, dulse could be an interesting alternative to *konbu*.

The data in Table 3.2, along with the findings for a series of other brown seaweeds from around the world (Mouritsen et al. 2019b), shows that none of them have any appreciably amounts of free glutamate and hence they have a very low umami potential. Even sugar kelp, which is of the same genus *Saccharina* as *konbu*, and oarweed, which is in the same order Laminariales as *konbu*, do not contain much glutamate. The same is true of all the investigated species of *Fucus*.

However, there is no simple relationship between the actual content of free glutamate in a *dashi* and a taster's perception of umami. This is demonstrated in Fig. 3.4 that shows the perceived umami intensity from a sensory analysis as a function of the amount of free glutamate in *dashi* prepared from different brown seaweeds. The quantitative sensory evaluation is based on a trained panel of tasters (Mouritsen et al. 2019b). Thus, even if a *dashi* is made of seaweeds that release very little free glutamate, it may by tasters be perceived to have a substantial umami taste.

3.4 Umami from Cephalopods

Cephalopods, specifically Neocoleoidea (squid, octopus, and cuttlefish), are used as marine food in most cuisines and food cultures across the world, particularly in coastal areas (Mouritsen and Styrbæk 2018a). Both the mantle, the arms and tentacles, the ink, and part of the intestines such as the hepatopancreas and female reproductive organs are used (cf. Fig. 3.3). Cephalopods are consumed raw as well as being prepared by boiling, steaming, frying, grilling, marinating, smoking, drying, and fermenting. Cephalopods are a good source of proteins, and there is a growing interest among chefs and gastroscientists to exploit marine food sources in a more diverse and sustainable fashion, e.g., to replace meat from land-animal production and other fisheries under pressure (Faxholm et al. 2018). The promises of this development of a cephalopod gastronomy are put in a particular perspective by the recent

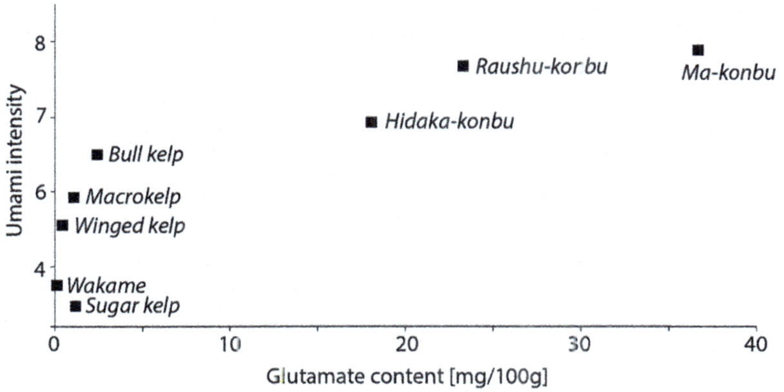

Fig. 3.3 Perceived umami intensity (scores on a scale 1–10) from a sensory analysis as a function of the amount of free glutamate in *dashi* prepared from different seaweeds. For Latin names, see Table 3.2. Macrokelp is *Macrocystis pyrifera*. Adapted from Mouritsen et al. (2019b)

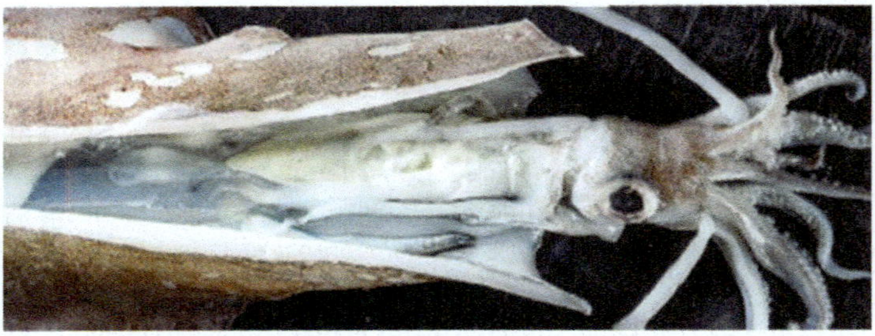

Fig. 3.4 Section of a squid (*Loligo forbesii*) where the mantle is cut open in order to expose parts of the interior, including the hepatopancreas, the gills, and the ink sack (Permission to print by courtesy of Jonas Drotner Mouritsen)

observation that the global populations of all types of cephalopods are on the rise (Doubleday et al. 2016).

In the present context it is of interest to reveal the umami potential of different cephalopods and their different parts. We have recently undertaken a series of studies of the gastrophysics and gastronomy of cephalopods (Mouritsen and Styrbæk 2018b) with a focus on texture and taste of Nordic squid (Faxholm et al. 2018), in particular umami (Fig. 3.4).

In the same way as other animals from salty waters, such as bonefish, shellfish, and seaweeds, cephalopods have many different tastes and flavours, but they share a common umami taste due to their high content of nucleic acids, like ATP, that enzymatically can be degraded into free nucleotides such as inosinate and adenylate under the proper conditions after the animal has been killed under not too stressed conditions. The simultaneous presence of both free nucleotides and free glutamate is the precondition for the very potent umami-synergy mechanism coming into play (Mouritsen and Khandelia 2012). Squid can contain high levels of adenylate, up to 184 mg/100 g (Yamaguchi and Ninomiya 2000), which is about as much as scallops and six times as much as a sun-ripe tomato.

We have recently measured the free amino acid contents of squid mantle from Nordic squid *Loligo forbesii* (Faxholm et al. 2018) and found a free glutamate content of 63 mg/100 g, somewhat larger than the value 20–30 mg/100 g reported for Japanese squid (UIC 2018). Although not specified, the latter value is likely to apply to Japanese flying squid (*Todarodes pacificus*). Compared to other types of meat, the free glutamate content in Nordic squid is substantially higher than beef (10 mg/100 g) and chicken (22 mg/100 g), but less than in scallop (140 mg/100 g) (Yamaguchi and Ninomiya 2000). Considering the relatively high amounts of free glutamate in raw squid, it is expected to have a rich umami taste, along with the sweet notes from other abundant free amino acids with sweet taste, such as alanine and glycine (Faxholm et al. 2018).

3.5 Umami and Sous Vide-Prepared Meat

Long-time preparation and ageing of meat products such as beef and pork are often considered a means to release umami compounds (Mouritsen and Styrbæk 2014). In recent years, long-time low-temperature cooking using sous vide techniques has gained renewed interest both in industrial preparations as well as in gastronomy and in the home kitchen (Baldwin 2012). Although a main focus has been on optimization of tenderness, juiciness, and flavour, the question has been raised regarding possible effects on umami taste, that is, release of free glutamate and nucleotides. Surprisingly few quantitative studies have been done along these lines on beef (Mortensen et al. 2015), and the major interest has concentrated on pork (Rotola-Pukkila et al. 2015) where it was found that free glutamate is released over time.

In a recent pilot study of beef tenderloin subjected to sous vide cooking (Clausen et al. 2018), both texture and free amino acid contents were measured under different cooking conditions (time, temperature). In order to determine the possible changes in total free amino acid content, the total glutamate contents of the samples, including both meat and juices, were assessed using different extraction techniques. An example of the resulting data for glutamate is reproduced in Fig. 3.5. Data for other temperatures and extraction techniques showed the same lack of time dependence: there is no increase of free glutamate content over time. The data for the other free amino acids show the same behaviour. Consequently, the umami taste of the sous vide-processed meat is the same as that of raw meat.

Figure 3.5 shows that the typical free glutamate content in the sous vide-cooked beef tenderloin samples is about 10 mg/100 g which is somewhat lower than was previously measured for beef meat, about 30 mg/100 g (Maga 1983; Ishiwatari et al. 2013). These variations are possibly caused by differences in the chosen meat cut and the extraction methods used.

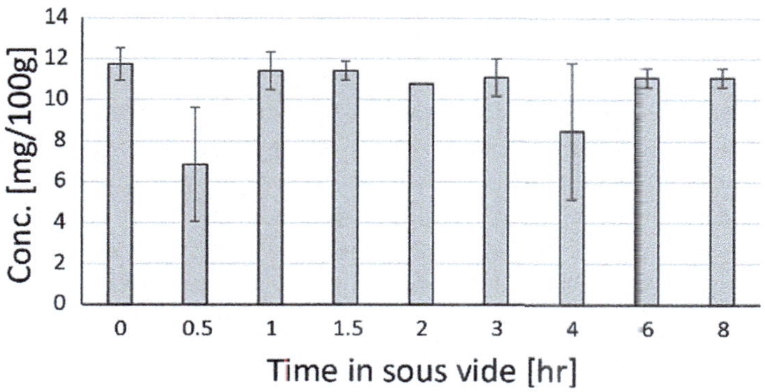

Fig. 3.5 Concentration of free glutamate in beef tenderloin prepared by sous vide techniques applied at 54 °C for a range of time periods. Reproduced from Clausen et al. (2018) with permission

Although the study in Fig. 3.5 implies that the total glutamate content does not change over time, it is quite possible that the free glutamate content in the released juice increases over time when the water-soluble glutamate seeps out from the cooking meat, as has been found in the case of pork (Rotola-Pukkila et al. 2015). Moreover, the results do not exclude that sous vide cooking can change the overall flavour of the meat and that sous vide techniques in fact can lead to more delicious and juicy beef. In the total flavour profile other taste compounds, fat, and aroma substances are likely to play a seminal role (Baldwin 2012), but from the point of view of pure glutamate, there is no increase of umami in the sous vide preparations of beef.

3.6 Umami and *Kokumi*

In connection with a study of flavour constituents in an aqueous extract of garlic in 1990, Ueda et al. (1990) observed that whereas this extract had little flavour of its own, it enhanced, even in very small amounts, the character of other flavours by imparting 'continuity, mouthfullness, and thickness' to the overall taste experience. It adds a so-called *koku* attribute and enhances food palatability by creating a special harmony of taste (Kuroda et al. 2004). Since none of the classical four basic tastes can elicit the same effect, and since when added to an umami solution the garlic extract also adds *koku*, Ueda et al. (1990) suggested the term *kokumi* to characterize this particular taste sensation. *Kokumi* substances have no taste themselves, but they influence basic tastes, e.g., enhancing salty, sweet, and umami, whereas they suppress bitterness. The effect on sour is still unclear. It was subsequently shown that certain tripeptides, such as glutathione (γ-Glu-Cys-Gly), are responsible for the *kokumi* flavour (Ueda et al. 1990). It is now known that the physiological mechanism behind the sensation is mediated by particular calcium-sensitive receptors in the taste buds (Maruyama et al. 2012). The mechanism behind *kokumi* sensation (i.e., sensation elicited by *kokumi* substances), however, is still only partly understood (Nishimura and Egusa 2016).

Nevertheless, there appears to be a particular interplay between umami and *kokumi*, even though most Westerners would have difficulty discerning and describing the difference between the two. Moreover, many kinds of, in particular fermented, foodstuff have both high amounts of umami compounds, such as free glutamate and nucleotides, as well as *kokumi* substances. A particular case is fermented fish sauces (Kuroda et al. 2012a; Miyamura et al. 2016) as described above. Gluthione has also been found in raw scallop and processed scallop products (Kuroda et al. 2012b), soy sauce (Kuroda et al. 2013), and fermented shrimp (Miyamura et al. 2014), products that are all rich in free adenylate, free glutamate, and free inosinate, respectively.

3.7 Public Dissemination and Outreach

The present project was carried out as part of the activities at the national Danish centre for taste studies and taste communication Smag for Livet (Taste for Life, www.smagforlivet.dk). The results were disseminated via a number of different activities, including the writing of popular science articles and books, scientific reviews, communication with chefs and fishermen via hands-on workshops, as well as production of teaching materials and courses for elementary school and high school children.

Acknowledgements This work was supported in part by a grant to Smag for Livet (Taste for Life) from Nordea-fonden.

References

Antony M, Blumenthal H, Bourdas A, Kinch D, Martinez V et al (2014) Umami: the fifth taste. Japan Publishing Trading Co, Tokyo

Baldwin DE (2012) Sous vide cooking: a review. Int J Gastron Food Sci 1:15–30

Clausen MP, Christensen M, Djurhuus TH, Duelund L, Mouritsen OG (2018) The quest for umami: can sous vide contribute? Int J Gastron Food Sci 13:129–133

Curtis RI (2009) Umami and the foods of classic antiquity. Am J Clin Nutr 90:712S–718S

Dalby A (2011) Geoponika. Prospect Books, Totnes

Doubleday ZA, Prowse TAA, Arkhipkin A, Pierce GJ, Semmens J, Steer M, Leporati SC, Lourenço S, Quetglas A, Sauer W, Gillanders BM (2016) Global proliferation of cephalopods. Curr Biol 26:R406–R407

Faxholm PL, Schmidt CV, Brønnum LB, Sun Y-T, Clausen MP (2018) Squids of the North: gastronomy and gastrophysics of Danish squid. Int J Gastron Food Sci 14:66–76

Gill TA (1990) Objective analysis of seafood quality. Food Rev Int 6:681–714

Grainger S (2010) Roman fish sauce. An experiment in archaeology. In: Procs of the Oxford symposium on Food and Cookery. Prospect Books, Devon, pp 121–131

Howgate P (2006) A review of the kinetics of degradation of inosine monophosphate in some species of fish during chilled storage. Int J Food Sci Technol 41:341–353

Ikeda I (2002) New seasonings. Chem Senses 27:847–849. [Translation from the original article in *J. Chem. Soc. Jpn.* 30:820-836 (1909)]

Ishiwatari N, Fukuoka M, Hamada-Sato N, Sakai N (2013) Decomposition kinetics of umami component during meat cooking. J Food Eng 119:324–331

Japanese Culinary Academy (2016) Flavor and seasonings: dashi, umami, and fermented foods. Shuhari Initiative Ltd, Tokyo

Katz SE (2012) The art of fermentation. Chelsea Green Publishing, Vermont

Komata Y (1990) Umami taste of seafoods. Food Rev Int 6:457–487

Kuroda M, Yamanaka T, Miyamura N (2004) Change in taste and flavour of food during aging with heating process. Generation of "KOKUMI" flavour during the heating of beef soup and beef extract. Jpn J Taste Smell Res 11:175–180

Kuroda M, Kato Y, Yamazaki J, Kai Y, Mizukoshi T, Miyano M, Eto Y (2012a) Determination and quantification of γ-glutamyl-valyl-glycine in commercial fish sauces. Agric Food Chem 60:7291–7296

Kuroda M, Kato Y, Yamazaki J, Kageyama N, Mizukoshi T, Miyano H, Eto Y (2012b) Determination of γ-glutamyl-valyl-glycine in raw scallop and processed scallop products using high pressure liquid chromatography-tandem mass spectrometry. Food Chem 134:1640–1644

Kuroda M, Kato Y, Yamazaki J, Eto Y (2013) Determination and quantification of the kokumi peptide, γ-glutamyl-valyl-glycine, in commercial soy sauces. Food Chem 141:823–828

Maga JA (1983) Flavor potentiators. CRC Crit Rev Food Sci Nutr 18:231–312

Maruyama Y, Yasuda R, Kuroda M, Eto Y (2012) Kokumi substances, enhancers of basic tastes, induce responses in calcium-sensing receptor expressing taste cells. PLoS One 7:e34489

Miyamura N, Kuroda M, Kato Y, Eto Y (2014) Determination and quantification of a kokumi peptide, 7-glutamyl-valyl-glycine, in fermented shrimp paste condiments. Food Sci Technol Res 20:699–703

Miyamura N, Kuroda M, Kato Y, Yamazaki J, Mizukoshi T, Miyano H (2016) Quantitative analysis of γ-glutamyl-valyl-glycine in fish sauces fermented with koji by LC/MS/MS. Chromatography 37:39–42

Mortensen LM, Frøst MB, Skibsted LH, Risbo J (2015) Long-time low-temperature cooking of beef: three dominant time-temperature behaviours of sensory properties. Flavour 4:2

Mouritsen OG (2013) Seaweeds. Edible, available & sustainable. Chicago University Press, Chicago

Mouritsen OG (2016) Deliciousness of food and a proper balance in fatty-acid composition as means to improve human health and regulate food intake. Flavour 5(1):1–13

Mouritsen OG (2017) Those taste weeds. J Appl Phycol 29:2159–2164

Mouritsen OG, Khandelia H (2012) Molecular mechanism of the allosteric enhancement of the umami taste sensation. FEBS J 279:3112–3120

Mouritsen OG, Styrbæk K (2014) Umami: unlocking the secrets of the fifth taste. Columbia University Press, New York

Mouritsen OG, Styrbæk K (2018a) Cephalopod gastronomy – a promise for the future. Front Comm Sci Environ Comm 3:38

Mouritsen OG, Styrbæk K (2018b) Blæksprutterne kommer – spis dem! Gyldendal, Copenhagen

Mouritsen OG, Williams L, Bjerregaard R, Duelund L (2012) Seaweeds for umami flavor in the New Nordic Cuisine. Flavour 1:4

Mouritsen OG, Dawczynski C, Duelund L, Jahreis G, Vetter W, Schroder M (2013) On the human consumption of the red seaweed dulse (*Palmaria palmata* (L.) Weber & Mohr). J Appl Phycol 25:1777–1791

Mouritsen OG, Calleja G, Duelund L, Frøst MB (2017) Flavour of fermented fish, insect, game, and pea sauces: garum revisited. Int J Gastron Food Sci 9:16–28

Mouritsen OG, Pérez-Lloréns JL, Rhatigan P (2019a) The rise of seaweed gastronomy: phycogastronomy. Bot Mar. 62:195-209

Mouritsen OG, Duelund L, Petersen MA, Hartmann AL, Frøst MB (2019b) Umami potential of brown seaweeds. J Appl Phycol. 31:1213–1232

Ninomiya K (1998) Natural occurrence. Food Rev Int 14:177–211

Nishimura T, Egusa A (2016) "Koku" involved in food palatability: an overview of pioneering work and outstanding questions. Kagaku to Seibutsu 54:102–108

O'Mahony M, Ishii R (1986) A comparison of English and Japanese taste languages: taste descriptive methodology, codability and the umami taste. Br J Psychol 77:161–174

Pérez-Lloréns JL, Hernández I, Vergara JJ, Brun FG, León À (2018) Those curious and delicious seaweeds. A fascinating voyage from biology to gastronomy. Servicio de Publicaciones de la Universidad de Cádiz, Cádiz

Rotola-Pukkil MK, Pihlajaviita ST, Mika T, Kaimainen MT, Hopia AI (2015) Concentration of umami compounds in pork meat and cooking juice with different cooking times and temperatures. J Food Sci 80:C2711–C2716

Teas J, Pino S, Critchley A, Braverman LE (2004) Variability of iodine content in common commercially available edible seaweeds. Thyroid 14:836–841

Ueda Y, Sakaguchi M, Hirayama K, Miyajima R, Kimizuka A (1990) Characteristic flavour constituents in water extract of garlic. Agric Biol Chem 54:163–169

UIC (2018) Umami Information Center, Tokyo. https://www.umamiinfo.com/richfood/foodstuff/seafood.php#ANCHOR08

Yamaguchi S, Ninomiya K (2000) Umami and food palatability. Am Soc Nutr Sci 130:921S–926S

Zhang FB, Klebansky B, Fine RM, Xu H, Pronin A, Liu H, Tachdjian C, Li X (2008) Molecular mechanism for the umami taste synergism. Proc Natl Acad Sci U S A 105:20930–20934

Ole G. Mouritsen, PhD DSc, is a biophysicist and professor of gastrophysics and culinary food innovation at the University of Copenhagen. He is director of the Danish national centre for taste, Smag for Livet, president of the Danish Gastronomical Academy, and appointed Japanese Cuisine Goodwill Ambassador. His fields of research include computational statistical physics, membrane biophysics, and culinary sciences, specifically gastrophysics. He is currently interested in exploring the use of macroalgae and cephalopods within gastronomy with a focus on particularly umami and texture/mouthfeel and how the insight can be used to prepare delicious vegetable dishes.

Chapter 4
Umami Compounds and Fats Involved in *Koku* Attribute of Pork Sausages

Toshihide Nishimura and Ai Egusa Saiga

Abstract This research was carried on in order to examine the role of umami compounds and fats in the sensation of "*koku* attributes" and to identify the aroma compounds contributing to the characteristic aroma in pork sausages.

Sensory analyses showed that the addition of umami compounds to sausage mix enhanced the intensity of retronasal aroma sensation, the complexity of flavor, the lingeringness of taste, and the intensity of umami of sausage. Then, sensory analyses also showed that addition of fat to sausage mix enhanced the mouthfulness and lingeringness of flavor as well as the intensity of umami taste in pork sausage. This result indicates that fat enhances the effect of "*koku* attribute" through the lingeringness of flavor. GC/MS analyses of aroma compounds released from pork sausages containing different amounts of fat have shown that β-pinene, 3-carene, D-limonene, octanal, nonanal, caryophyllene, and methyl eugenol were identified in all these sausages. The amounts these aroma compounds released from sausage have been shown to decrease by addition of fat into sausage, indicating that fat was able to hold aroma compounds in sausages. From these results, the lingeringness effect of aroma sensation in "*koku* attribute" of sausage was shown to be enhanced by addition of fat.

Keywords Umami compound · Fat · *Koku* attribute · Sausage · Aroma compounds

T. Nishimura (✉)
Faculty of Nutrition, Kagawa Nutrition University, Sakado, Saitama, Japan
e-mail: nishimura.toshihide@eiyo.ac.jp

A. E. Saiga
Faculty of Applied Biological Science, Nippon Veterinary and Life Science University, Musashino-City, Tokyo, Japan

T. Nishimura, M. Kuroda (eds.), *Koku in Food Science and Physiology*,
https://doi.org/10.1007/978-981-13-8453-0_4

4.1 Introduction

Factors such as taste, aroma, texture (tenderness, viscosity, smoothness, or juiciness), color, temperature, and shape are involved in food palatability. Among many factors, in general, taste is caused by taste compounds in foods which are water-soluble. Aroma is caused by aroma compounds which are volatile compounds released from foods. Aroma is classified into orthonasal and retronasal aromas, and the latter aroma is sensed after foods are put into our mouth. This retronasal aroma is thought to be the most important factor involved in food palatability for us. There is also found to be interaction between taste sensation and retronasal aroma sensation in our sensations.

Recently, we proposed that *koku* attribute in foods is one of the objective factors involved in their palatability, and the total sensation of complexity, mouthfulness, and lingeringness is caused by factors such as taste, aroma, and texture. In general, foods such as curry, stew, ramen noodles, and natural cheese which are heated, conditioned, or fermented for a long time give us *koku* attributes such as complexity, mouthfulness, and lingering. There is the objective intensity like a scale in complexity, mouthfulness, and lingeringness in "*koku* attribute," and their intensity depends on the amount of stimulations by taste compounds, aroma compounds, and texture compounds. Compounds contributing to "*koku* attributes" and the suitable intensity of *koku* attributes are different depending on foods. Research on these compounds and intensity involved in attribute in each food is needed from now.

The mouthfulness of *koku* attributes of foods such as curry and stew is sensed by the spread of flavor in the mouth. This spread of flavor is thought to be caused by umami compounds. The miso soup (50g/L) without seasonings showed us very weak sensation of mouthfulness in the space of our oral cavity, while it has the characteristic flavor and complexity of miso soup. The addition of umami compounds to this miso soup can enlarge the intensity of spread of flavor sensation in our oral cavity. Yamaguchi et al. have reported that umami compounds had the effect of flavor enhancer of foods when added to foods (Yamaguchi and Kimizuka 1979). Recently, Nishimura et al. examined the mechanism of flavor enhancer by the addition of umami compounds to foods. The analysis using model chicken extract clarified that the addition of monosodium glutamate to the extract made the sensation of retronasal aroma 2.5 times higher than no addition (Nishimura et al. 2016a). This effect became higher, and the concentration of added MSG was increased to 0.3%. The addition of MSG at over 0.3% made the retronasal aroma of the extract weaker, while the intensity of umami taste became stronger. Therefore, umami compounds have a strong effect on the enhancement of mouthfulness in *koku* attributes by enhancing flavor, especially retronasal aroma.

When umami compounds are put into our mouth, the tactile stimulation of the tongue continued for long time. Although the sulfur-containing compounds or *kokumi* peptides also show the effect of lingeringness, there are few evidences showing that only these compounds have the same effect as umami compounds.

Another lingeringness in *koku* attribute is caused by the holding activity of aroma compounds with lipids. Ramen noodles in pork bone soup and marbled beef of Japanese black cattle are rich dishes because of containing lipid or fat in these foods. Although pure lipids or fats have no taste and aroma, we can sense complex taste and aroma when we put cooked lipids or fats into our mouth. Nishimura et al. (2016b) clarified that this phenomenon was brought about by taste and aroma compounds bound to lipids and fats nonspecifically. They have found that a precipitate of heat-treated onion concentrates (HOC) interacted with aroma compounds and thus enhanced aroma persistence including lingering of aroma as a *koku* attribute of consommé. The key compounds in the precipitate were the phytosterols, that is, β-sitosterol and stigmasterol. To our knowledge, this is the first study to show that phytosterols can interact with aroma compounds in foods. Lipids or fats in foods contribute to the complexity and lingeringness of *koku* attribute in foods. It is well known that the presence of lipids influences the persistence of aroma compounds in foods. Aroma compounds in foods partially bind to lipids or fats by hydrophobic interaction.

Pork sausage is very popular and considered as one of the delicious foods. However, the clarification of factors involved in its deliciousness has not been known yet. Pork sausages are produced from minced pork meat by addition of fats, curing agents, and some seasonings. Especially, the addition of umami compounds and fats to sausage mix improved the palatability of pork sausages. There are few reports examining the effect of addition of these compounds on the quality of sausages. Recently, umami compounds and fats are also found to be involved in *koku* attributes.

So, this research was carried out in order to elucidate the role of umami compounds and fats in the sensation of "*koku* attributes", and to identify the aroma compounds contributing to the characteristic aroma in pork sausages.

4.2 Materials and Methods

4.2.1 Preparation of Samples

The sausages with addition of a regular amount of umami compounds, 50% amount of the regular one, 10% amount of the regular one, and without addition of umami compounds were prepared for the experiment concerning the effect of umami compounds. In another experiment, sausages with addition of a regular amount of fat, 50% amount of the regular one, and without addition of fat were also prepared for the experiment concerning the effect of fat. These sausages were heated in hot water at 90 °C until the temperature in the center of sausages reached to 50 °C.

4.2.2 Sensory Evaluation

4.2.2.1 Effect of the Addition of Umami Compounds on the Quality of Sausages

Sensory evaluation for the following 7 items was conducted in order to examine the effect of addition of umami compounds to sausages. The seven items were intensity of umami taste, lingeringness of taste, complexity of flavor, mouthfulness of flavor, retronasal aroma, mouthfulness of aroma, and lingeringness of aroma. The sausages with addition of a regular amount of umami compounds were used as standard for evaluation. The sausages with 50% amount of the regular one, with 10% amount of the regular one, and without addition of umami compounds were evaluated in the difference of 7 items compared to standard one. Trained panelists conducted sensory evaluation at room temperature of 25 °C. Four different kinds of sausages were provided and evaluated for each piece of sausages according to the scale. Mineral water was also provided to cleanse the mouth. To analyze the results of the sensory evaluation, two-way analysis of variance and Tukey's test were used.

4.2.2.2 Effect of the Addition of Fats on the Quality of Sausages

Sensory evaluation for the following 5 items was conducted in order to examine the effect of addition of fat to sausages. Five items were the intensity of spicy aroma, intensity of umami taste, intensity of saltiness, mouthfulness of flavor, and lingeringness of flavor. The score from −3 (very weak) to 3 (very strong) was used to evaluate each item of samples. Fifteen panelists conducted sensory evaluation at room temperature of 25 °C. One sausage was cut into two pieces. Then, the pieces containing different amounts of fat were provided to panelists and evaluated according to the five items. Panelists evaluated the intensity of flavor after chewing 10 times for mouthfulness and evaluated the intensity of lingering flavor after chewing 10 times and swallowing them. In the case of the evaluation of the intensity of umami taste or saltiness, panelists evaluate them while clipping their nose. The warmed mineral water was also provided to cleanse the palate. To analyze the results of the sensory evaluation, two-way analysis of variance and Tukey's test were used.

4.2.2.3 Panelists

Fifteen panelists were trained for at least 2 months by sensory evaluation using discrimination test among five basic taste solutions, among different intensities of aroma in the aroma standard.

4.2.2.4 Analyses of Aroma Compounds

1. Extraction of aroma compounds from sausages with diethyl-ether

Twenty milliliter of diethyl ether was added to 5 grams of minced sausages, and aroma compounds were extracted at 2 °C for 16 h.

2. Trap of aroma compounds released from sausages in head space (HD)

Five grams of minced sausages were put into 40-ml vial tube, and Mono Trap (GL- Science, Tokyo, Japan) was put into the vial. This vial was put in a chamber and heated at 60 °C for 2 h. The aroma compounds released from sausages were collected in head space (HS) of sausages with the Mono Trap which is a new hybrid adsorptive. Aroma compounds were extracted from Mono Trap with 200 microliters of diethyl-ether and exposed to ultrasound for 5 min.

3. GC/MS

The aroma compounds collected from each sausage by diethyl-ether or HS were analyzed by GC/MS and identified as follows: GC was performed on an Agilent 7890A GC System (Agilent Technologies, USA) equipped with a CP-wax 52 CB column (GL Science Co., Tokyo, Japan) of 0.25 mm i.d. and 30-m-long, Agilent 7693 Autosampler, and Agilent Technologies 5975C inert MSD with Triple-Axis Detector. An initial oven temperature of 40 °C for 5 min was used, the temperature was then increased to 10 °C/min at 240 °C.

4. GC-O

Characteristic aroma compounds collected from each sausage by HS were analyzed by GC/O, and the aroma compound contributing to sausage flavor was identified by AEDA (aroma extract dilution analysis). GC was performed on a GC-2014 (Shimadzu Co., Kyoto, Japan) equipped with a TC-Wax capillary column (GL Sciences Inc., Tokyo, Japan) of 0.25 mm i.d. and 30 m long, with an FID at 260 °C. An initial oven temperature of 40 °C for 4 min was used; the temperature was then increased to 6 °C/min at 240 °C.

4.3 Results and Discussion

4.3.1 Effect of Amount of Umami Compounds on Sensory Characteristics of Pork Sausages

We evaluated characteristics in flavor of sausages containing different amounts of umami compounds by sensory evaluation for seven sensation items such as intensity of retronasal aroma, mouthfulness of aroma, lingeringness of aroma, intensity of umami taste, lingeringness of taste, complexity of flavor, and mouthfulness of flavor.

The addition of umami compounds to sausages enhanced the complexity of flavor, lingeringness of flavor, and intensity of retronasal aroma sensation (Fig. 4.1). Although the addition of umami compounds to the model chicken soup has already been shown to enhance the intensity of its retronasal aroma sensation, it was the first report that the addition of umami compounds into sausage enhanced the complexity of flavor, lingeringness of flavor, and intensity of retronasal aroma sensation.

Yamaguchi et al. have reported that umami compounds had a flavor enhancer effect when they were added to foods (Yamaguchi and Kimizuka 1979). Recently, Nishimura et al. have also found that the addition of monosodium glutamate to the model chicken extract enhanced the intensity of sensation of retronasal aroma until 2.5 times higher than no addition (Nishimura et al. 2016a). Therefore, the flavor of foods containing umami compounds was enhanced in mouthfulness without

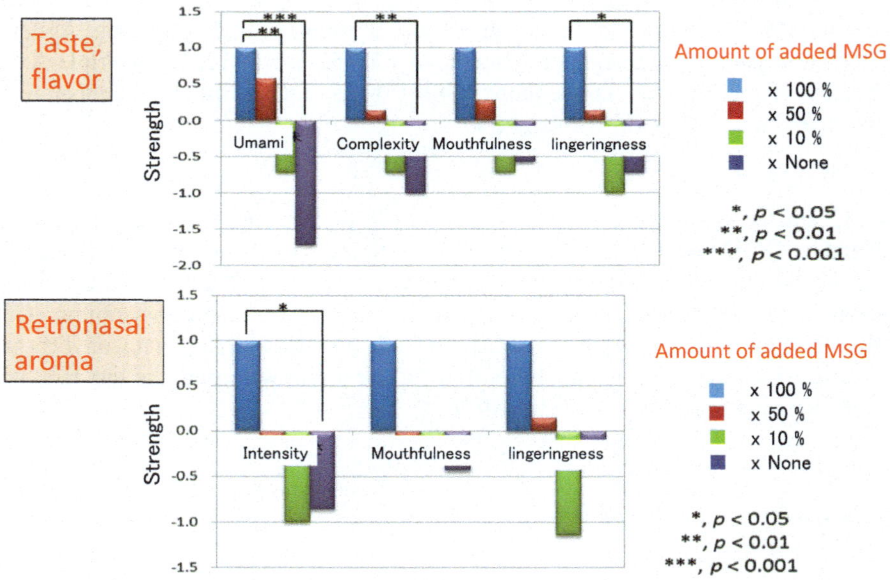

Fig. 4.1 Effect of addition of umami compounds on enhancement of flavor in sausages

changing the characteristics of their flavor by *koku*-enhancing effect of umami compounds.

4.3.2 *Effect of Fats Contents on Sensory Characteristics of Pork Sausages*

The sensory characteristics of three kinds of pork sausages containing different amounts of fats were examined by sensory evaluation for five sensation items such as intensity of spicy aroma, intensity of umami taste, intensity of saltiness, mouthfulness of flavor, and lingeringness of flavor.

Sensory analyses showed that addition of fat enhanced the intensity of the mouthfulness and lingeringness of flavor in pork sausage as well as the intensity of umami taste (Fig. 4.2). This result indicated that the addition of fats enhances the "*koku* attribute" by the lingering aroma sensation. In our previous report, phytosterols such as β-sitosterol and stigmasterol were shown to be clarified to hold aroma compounds and to enhance the lingeringness of aroma in *koku* attribute of consommé soup. In this research, fats enhance the lingeringness of flavor in *koku* attributes. The next problem to be resolved is to clarify what kind of fat has the strongest effect on *koku* attribute. On the other hand, there is no effect of addition of fat on the enhancement of the intensity of spicy aroma sensation in pork sausage.

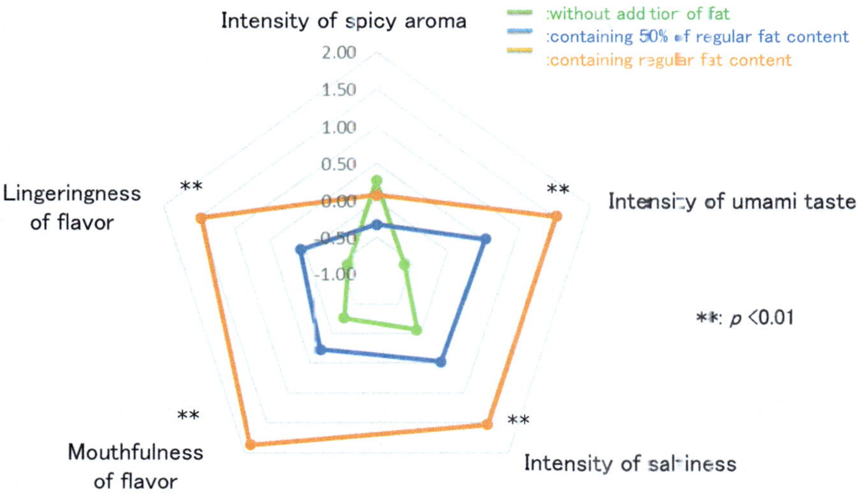

Fig. 4.2 Difference in sensory quality among three kinds of pork sausages containing different amounts of fat

4.3.3 Analyses and Identification of Aroma Compounds Involved in Non-smoked Sausages

GC/MS analyses of aroma compounds extracted from pork sausages showed that 13 compounds, such as β-pinene, 3-carene, D-limonene, acetic acid, copaene, caryophyllene, methyl eugenol, 6-octadecanoic acid, 2-octyl-cycropropanoic acid, oleic acid, cis-vaccenic acid, 2-methoxy-3-(2-propenyl)-phenol, and (E)-9-octadecenoic acid, were identified. These compounds seem to be derived from spices added to sausages in the production process.

Next, AEDA analyses with GC/O were carried out to specify the aroma compounds involved in characteristic of non-smoked sausages from 13 aroma compounds. The aroma of β-pinene, 3-carene, D-limonene, acetic acid, copaene, and methyl eugenol was detected even in 2048 diluted sample, and these 6 compounds seemed to be involved in the characteristic of non-smoked pork sausages (Figs. 4.3 and Table 4.1).

4.3.4 Effect of Fat Content on the Amount of Released Aroma Compounds

In the case of GC/MS analyses of aroma compounds released into HS from pork sausages containing different amounts of fats, 7 aroma compounds, such as β-pinene, 3-carene, D-limonene, octanal, nonanal, caryophyllene, and methyl eugenol, were identified in these sausages (Fig. 4.4). The amount of these aroma compounds released from sausage has been shown to decrease by addition of fat into sausage, indicating that fat was able to hold aroma compounds in sausages. However, these aroma compounds bound to fats seemed to come off gradually after putting sausages into the oral cavity and chewing them (Fig. 4.5). From these results, the lingeringness effect of aroma sensation in "*koku* attribute" of sausage was also shown to be enhanced by addition of fat. The released aroma compounds from sausages containing fats seems to be larger than that from sausage without addition of fat in chewing sausages (Fig. 4.6).

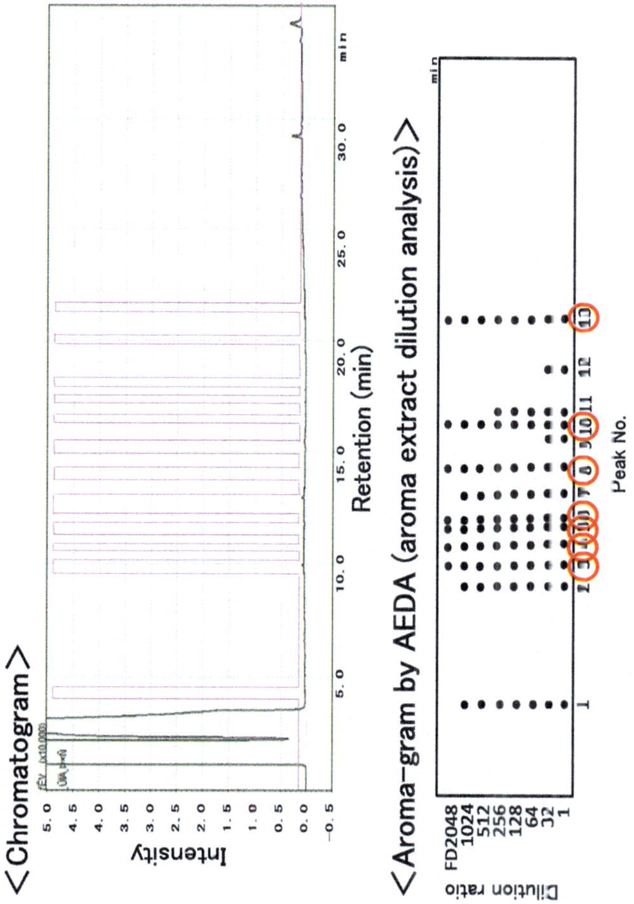

Fig. 4.3 Chromatogram and aroma-gram in GC/O analyses

Table 4.1 Candidate aroma compounds detected in non-smoked sausages

Peak No.	Characteristics of aroma	Candidate compounds
3	mushroom, wood, beans	β-pinene
4	egg soup, green-smelling	3-carene
5	sour, green-smelling	D-limonene
6	sour, vinegar	acetic acid
8	dust, rubber	un-identified
10	cucumber, green-smelling	copaene
13	spice, curry, herbalmedicine	methyl eugenol

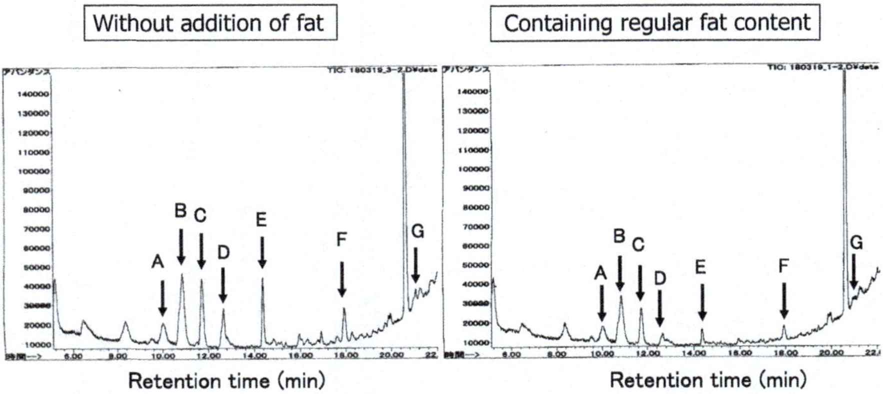

Fig. 4.4 GC-chromatogram of aroma compounds released from non-smoked pork sausage. [Compounds] A : β-Pinene, B : 3-Carene, C : D-Limonene, D : Octanal, E : Nonanal, F: Caryophyllene, G : Methyl eugenol

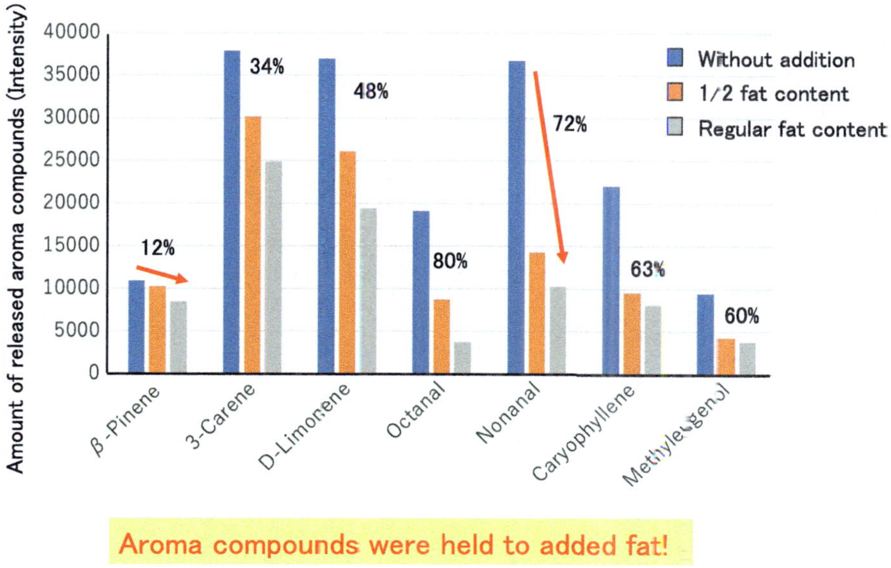

Fig. 4.5 Effect of fat content on the released aroma compounds

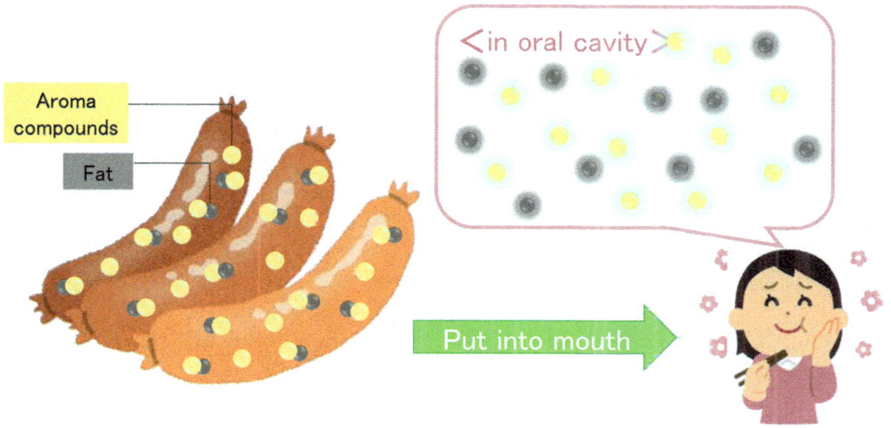

Fig. 4.6 Mechanisms for larger release of aroma compounds from sausages containing fat

References

Nishimura T, Goto S, Miura K, Takakura Y, Egusa AS, Wakabayashi H (2016a) Umami compounds enhance the intensity of retronasal sensation of aromas from model chicken soups. Food Chem 196:577–583

Nishimura T, Egusa AS, Nagao A, Odahara T, Sugise T, Mizoguchi N, Nosho Y (2016b) Phytosterols in onion contribute to a sensation of lingering of aroma, a *koku* attribute. Food Chem 192:724–728

Yamaguchi S, Kimizuka A (1979) Psychometric studies on the taste of monosodium glutamate. In: Filer LJ Jr et al (eds) Glutamic acid: advances biochemistry and physiology. Raven Press, New York, pp 35–54

Chapter 5
The Components Contributing to the Thickness of Beer Aroma

Toru Kishimoto

Abstract This study investigated how aroma components contribute to the overall aroma of Pilsner-type beer through the construction of aroma simulation models, using 76 odor-active components. In the process of the current study, the components contributing to the thickness of beer aroma, which were supposed to result in "*koku*," were investigated.

Gas chromatography-olfactometry (GC-O) analyses of the aroma components extracted from Pilsner-type beer were performed. The concentrations of all 76 components, which were identified after the GC-O analyses, were accurately quantitated. The odorants were then used in the following aroma simulation experiments, with ethanol and water as matrices.

On the first process of the current study, the odorants with higher odor intensity in GC-O analysis and odor activity values (OAVs) were selected at first, and the selected 25 odorants were used in an aroma simulation experiment. The results of sensory analysis showed that the simulation model using 25 odorants exhibited unbalanced characteristics and lacked the thickness (the total amount of aroma), malty/cereal and estery characteristics, and the similarity compared to the reference Pilsner-type beer.

On the next step, the components contributing to malty/cereal characteristics that construct the frame structure of beer aroma, related to the thickness of beer aroma, were analyzed, through the fractionations of whole beer aroma. The 24 odorants, which were assumed to complement the malty/cereal characteristic and thickness, were identified. Addition of the 24 odorants could not improve the aroma quality of the simulation model. On the final process of the study, other remaining 27 insignificant odorants without independent contributions, not reminiscent of beer aroma, were added to the recombinant. The addition of these odorants dramatically improved the quality of the recombinants.

The aroma simulation experiments using 76 odorants indicated that the synergistic contributions of the multiple insignificant odorants, with no independent contri-

T. Kishimoto (✉)
Research Laboratories for Alcohol Beverages, Asahi Breweries Ltd, Moriya-City, Ibaraki, Japan
e-mail: kishimoto.toru.x76@kyoto-u.jp

T. Nishimura, M. Kuroda (eds.), *Koku in Food Science and Physiology*,
https://doi.org/10.1007/978-981-13-8453-0_5

butions to the overall aroma characteristics, were necessary to construct the frame structure and the thickness of beer aroma.

Keywords Beer aroma · Thickness · Gas chromatography-olfactometry · GC-O · Aroma simulation · OAV · Threshold · *Koku*

5.1 Introduction

"*Koku*" is an indispensable feature to discuss the taste of beer. In market research on beer, "*koku*" and "*kire*" (crispness) are inevitable attributes in evaluating beer taste, and the palatability was expressed by a multiple regression model using "*koku*" and "*kire*" (Kaneda et al. 2002).

In the academic societies overseas related to beer, "*koku*" is expressed as "body" and defined as fullness of flavor and mouthfeel (Sensory analysis 1992). *Koku*, body, could be induced by the integration of basic taste, odor, tactile sense, and texture stimuli in beer (Kaneda et al. 2002). In past researches of beer taste, contributions of bitterness, astringent components, dextrin, and low-molecular-weight polypeptides were discussed, though the contributions of aroma to the "*koku*" of beer were not investigated.

This study investigated how aroma components contribute to the overall aroma of Pilsner-type beer through the construction of aroma simulation models, using 76 odor-active components (Kishimoto et al. 2018). In the process of the current study, the components contributing to the thickness of beer aroma, which were supposed to result in "*koku*," were investigated.

Many studies have been performed to reveal the odorants that contribute to the beer aroma, using gas chromatography-olfactometry (GC-O) analysis (Mayol and Acree 2001; Blank 2002) and by evaluating their odor activity value (OAV), which is the ratio of a component's concentration to its odor threshold (Schieberle 1995; Grosch 2001). When discussing the savory characteristics of foods, the contributing components are complicated and the contributions of subthreshold components have been reported. Kurobayashi et al. investigated the flavor-enhancing effects of volatile constituents at subthreshold levels in celery (Kurobayashi et al. 2008). In their research, the samples were prepared by adding celery fractions to chicken broth at a concentration such that their distinct odors were not detected. They found that the volatile compounds in celery enhanced the complex flavor, including "*koku*," of chicken broth more than the nonvolatile compounds. The results of the sensory evaluation revealed that sedanenolide, 3-n-butylphthalide, and sedanolide contribute to the complex flavor of chicken broth, and sedanenolide was most effective. In an investigation by Miyazawa et al. (2008), they used a mixture including maple lactone and carboxylic acids in the sensory difference test in order to quantify the interaction, where the stimuli were controlled by air-dilution olfactometry. They found that the addition of subthreshold carboxylic acids increased the statistical probability of identification in the test, and it was revealed that a positive interaction (synergistic effect) occurred between the odorants at perithreshold concentrations in the mixture.

In the current study (Kishimoto et al. 2018), the aroma simulation experiments of Pilsner-type beer were conducted using the odorants at the concentrations as they were in the reference beer. In particular, the contributions of the odorants, with no independent contributions to the overall aroma characteristics, were examined.

5.2 Results and Discussion

The details of analysis methods, chemicals, were described in the original manuscripts (Kishimoto et al. 2018). A commercial Pilsner-type beer containing 5.2% (v/v) alcohol, with a bitterness unit (European Brewery Convention 2010) of 20.0 B.U., obtained from the Ibaraki plant of Asahi Breweries, Ltd. (Ibaraki, Japan), immediately after production, was used as the reference beer. Within the frame of the current study, this commercial Pilsner-type beer was assigned as the reference beer and was examined throughout the experiments.

5.2.1 *Preparation of the Extracts for GC-O Analysis*

The CH_2Cl_2 extracts of the odorants were distilled using the SAFE apparatus to prevent the formation of artifacts from nonvolatile compounds in the GC-O inlets (Cabrita et al. 2012). The extract was carefully concentrated using a Kuderna-Danish evaporative concentrator in order to reduce the loss of highly volatile odorants. It was checked that the original aroma characteristics were recovered in the concentrate of extract P. CharmAnalysis™ was used in GC-O analyses of this study, because the aroma components are carried under the airflow at 20 L/min (Acree and Barnard 1996); thus, the flow of the odorants do not stay at the sniff port and the boundaries between the aroma components are clearly defined in the CharmAnalysis™. The aroma intensities were integrated to yield the Charm values (Acree 1993).

5.2.2 *Identification and Quantitation of Odorants*

Sixty-five odor-active compounds were detected in a GC-O analysis of whole aroma extract of the reference beer (extract P), and 56 odorants were identified by comparing their odor qualities, RIs, and mass spectra with those of authentic compounds on a DB-WAX column. In the following experiment to examine the odorants contributing to malty/cereal characteristics, whole aroma extract (extract Q) was fractionated. The GC-O analyses of the entire fractions of extract Q detected 87 odor-active compounds and identified 66 compounds. A total of 76 odor-active components were identified after the analysis of extract P and the entire fractions of extract Q.

Of the 76 odorants, 10 components (3-methylbutanal, hexanal, methional, (E)-2-nonenal, (E,Z)-2,6-nonadienal, phenylacetaldehyde, (E,E)-2,4-decadienal, β-ionone, trans-4,5-epoxy-(E)-2-decenal, 2-aminoacetophenone) were identified only in extract P, and 20 components (2-methylbutanal, ethyl propionate, 2,3-butanedione, 2,3-pentanedione, 2-methyl-1-propanol, ethyl 4-methylpentanoate, dimethyl trisulfide, 2-sulfanylethyl acetate, furfural, (E,E)-2,4-heptadienal, benzaldehyde, benzyl mercaptan, 3-sulfanyl-3-methyl-1-butanol, γ-hexalactone, citronellol, (E,E)-2,4-decadienal, cyclotene, γ-octalactone, ethyl cinnamate, maple furanone) were identified only in the entire fractions of extract Q.

The accurate concentrations of 76 identified odorants were quantitated by using multiple analytical methods, mainly via the SIDA method using GC/MS system or GC triple quadrupole MS system. The GC/MS systems were not used for the quantitation of 4-vinylguaiacol, 4-vinylphenol, and vanillin, since these odorants were supposed to be synthesized artificially from certain acids in the high-temperature GC inlets (Fiddler et al. 1967; Arrieta-Baez et al. 2012), and actually, the increases of these compounds after an injection into the GC apparatus were confirmed (data not shown). Concentrations of the 76 odorants were described in Table 5.1.

5.2.3 Sensory Evaluation

In the discussion among the panelists, the sensory characteristics of "malty/cereal" aroma, "fusel alcohol-like" aroma, "estery" aroma, and the "thickness (total amount of aroma)" were specified for aroma recombinants to be improved and then were evaluated in the sensory tests, with the evaluation scales ranging from −3.00 to +3.00, where the value of the reference beer was set at 0.00. The similarity was also evaluated by comparison with the reference beer, whose score was set at 100%.

The hoppy flavor was not evaluated in the sensory analyses of recombinants within the frame of this study, since the hoppy flavor was not characterized for the reference Pilsner-type beer and lower levels of reported hop-derived odorants (Kishimoto et al. 2005, 2006, 2007, 2008) were contained in the reference beer.

5.2.4 Construction of the Base Framework Structure Using 25 Odorants with Higher Charm Values and OAVs

Of the 65 odorants that were identified after the analysis of extract P, the 38 odorants with Charm values higher than 500 (Kishimoto et al. 2018) were prescreened, and the threshold values in the reference beer were determined for these 38 odorants. It was revealed that the OAVs of only 9 odorants exceeded 1.0. To examine the

Table 5.1 76 odor-active compounds used in the construction of the aroma recombinants of Pilsner-type beer (Kishimoto et al. 2018)

Component group	Odor quality	Identified compound	CAS no.	Concentration (μg/L)	Differential threshold value (μg/L) in beer	Independent contributions to overall aroma	RIs on DB-WAX column
A	Pineapple	Ethyl acetate	141-78-6	15300	21000	–	891
A	Potato almond	3-Methyl butanol	590-86-3	4.9	9.6	–	915
A	Banana	Isoamyl acetate	123-92-2	1230	724	+	1111
A	Fusel alcohol	3-Methyl-1-butanol	123-51-3	59600	16800	+	1215
A	Sweet, fruit	Ethyl hexanoate	123-66-0	119.2	163	–	1235
A	Nut, mushroom	1-Octen-3-one	4312-99-6	0.0066	0.0026	–	1296
A	Nut, thiamine	2-Methyl-3-furanthiol	28588-74-1	0.075	0.19	–	1317
A	Metallic, raw fish	(Z)-1, 5-Octadien-3-one	65767-22-8	0.00014	0.0004	–	1379
A	Ester, fruity	Ethyl octanoate	106-32-1	159.5	290	–	1440
A	Potato, soy sauce	Methional	3268-49-3	1.2	1.8	–	1453
A	Cardboard	(E)-2-Nonenal	2463-53-8	0.082	0.1	–	1537
A	Floral	(R)-Linalool	126-91-0	1.2	1	–	1546
A	Cooked onion	2-Sulfanyl-3-methyl-1-butanol	116229-37-9	0.27	0.29	–	1640
A	Rancid, cheese	3-Methyl butanoic acid	503-74-2	454.9	1230	–	1676
A	Soy sauce, fusel	Methionol	505-10-2	822	1397	–	1710
A	Fruity, catty,	3-Sulfanylhexyl acetate	136954-20-6	0.0047	0.005	–	1730

(continued)

Table 5.1 (continued)

Component group	Odor quality	Identified compound	CAS no.	Concentration (μg/L)	Differential threshold value (μg/L) in beer	Independent contributions to overall aroma	RIs on DB-WAX column
A	Strawberry, rose, honey	(E)-β-Damascenone	23696-85-7	1.8	2.5	–	1825
A	Catty, fruity	3-Sulfany1-1-hexanol	51755-83-0	0.044	0.055	–	1847
A	Floral, rose	Geraniol	106-24-1	2.7	7	–	1854
A	Floral, fusel	2-Phenylethyl alcohol	1960/12/8	27669	7739	+	1910
A	Sweet, milk	γ-Nonalactone	104-61-0	33.7	11.2	+	2035
A	Caramel, sour-sweet	Furaneol	3658-77-3	290.6	112	+	2040
A	Caramel, burnt sugar	Sotolon	28664-35-9	1.5	0.54	+	2195
A	Smoke, sweet, phenolic	4-Vinylguaiacol	7786-61-0	116.6	98	–	2210
A	Foxy, grape	2-Aminoacetophenone	551-93-9	2.4	5	–	2228
B	Potato almond	2-Methyl butanol	96-17-3	1.7		–	910
B	Roasted, skunky	3-Methy1-2-butene-1-thiol	5287-45-6	0.0029	0.007	–	1105
B	Cereal	2-Acety1-1-pyrroline	85213-22-5	0.54	3	–	1338
B	Earth, nut	2,3,5-Trimethylpyrazine	14667-55-1	0.24		–	1411
B	Rubber	2-Sulfanyl ethyl acetate	5862-40-8	2.3	6.3	–	1438
B	Bread, almond, sweet	Furfural	1998/1/1	20.4		–	1465

B	Earth, nut	2-Ethyl-3,5(6)-dimethylpyrazine	13925-07-0	0.25		–	1470
B	Almond, burnt sugar	Benzaldehyde	100-52-7	1.6		–	1530
B	Roast, burnt	Benzyl mercaptan	100-53-8	0.0061		–	1615
B	Coconut	γ-Hexalactone	695-06-7	22.6		–	1724
B	Caramel, burnt sugar	Cyclotene	80-71-7	20.4		–	1830
B	Smoke, sweet, phenolic	Guaiacol	1990/5/1	3	65	–	1864
B	Mushroom, cinnamon	Ethyl 3-phenylpropionate	2021-28-5	1.6	5		1897
B	Coconut, sweet	γ-Octalactone	104-50-7	3.7		–	1925
B	Caramel	Maltol	118-71-8	1243		–	1981
B	Caramel, burnt sugar	Homofuraneol	27538-09-6	11.4	68	–	2090
B	Mushroom, cinnamon	Ethyl cinnamate	103-36-6	0.93	2.4	–	2135
B	Caramel, burnt sugar	Maple furanone	698-10-2	0.13		–	2263
B	Smoke, sweet, phenolic	4-Vinylphenol	2628-17-3	78.9		–	2380
B	Mothball, fecal	Indole	120-72-9	0.72		–	2444
B	Mothball, fecal	3-Methylindole	83-34-1	0.054		–	2490
B	Vanilla, coconut	Vanillin	121-33-5	3	48	–	2585
B	Mushroom, cinnamon	3-Phenylpropionic acid	501-52-0	20.2		–	2630

(continued)

Table 5.1 (continued)

Component group	Odor quality	Identified compound	CAS no.	Concentration (μg/L)	Differential threshold value (μg/L) in beer	Independent contributions to overall aroma	RIs on DB-WAX column
B	Raspberry	Raspberry ketone	5471-51-2	5	21.2	–	2978
C	Fruit, apple	Ethyl propionate	105-37-3	96.2		–	952
C	Citrus, apple-like	Ethyl 2-methyl propanoate	97-62-1	0.91	6.3	–	965
C	Butter	2,3-Butanedione	431-03-8	13		–	970
C	Fruit, apple	Ethyl butanoate	105-54-4	73.1	367	–	1036
C	Butter	2,3-Pentanedione	600-14-6	5		–	1052
C	Fruit, pineapple	Ethyl 2-methyl butanoate	7452-79-1	0.21	1.1	–	1055
C	Citrus, pineapple	Ethyl 3-methyl butanoate	108-64-5	0.41	2	–	1065
C	Grass, green	Hexanal	66-25-1	6.2		–	1090
C	Sweet, musty	2-Methy1-1-propanol	78-83-1	12400		–	1097
C	Fruit, apple-like	Ethyl 4-methyl pentanoate	25415-67-2	0.14	1	–	1198
C	Sulfur, cabbage	Dimethyl trisulfide	3658-80-8	0.00012	0.016	–	1391
C	Fresh, orange	Decanal	112-31-2	1.5		–	1498
C	Fresh, cucumber, fried	(E,E)-2,4-Heptadienal	4313/3/5	2.7		–	1503
C	Cucumber, green	(E,Z)-2,6-Nonadienal	557-48-2	0.012		–	1590
C	Rancid, cheese	Butanoic acid	107-92-6	440	1899	–	1621

C	Rose, floral	Phenylacetaldehyde	122-78-1	4.5	39.7	–	1646
C	Cooked onion	3-Sulfanyl-3-methyl-1-butanol	34300-94-2	0.33		–	1658
C	Citrus, floral, rose	Citronellol	106-22-9	2.3	7.5	–	1770
C	Fat, fried	(E,E)-2,4-Decadienal	25152-84-5	0.032		–	1815
C	Floral, mint	2-Phenylethyl acetate	103-45-7	382.5	2760	–	1820
C	Rancid, sweaty	Hexanoic acid	142-62-1	910		–	1849
C	Floral, violet	β-Ionone	14901-07-6	0.013	0.6	–	1954
C	Metallic, raw fish	trans 4,5 Epoxy-(E)-2-decenal	134454-31-2	0.0037	0.059	–	2020
C	Sweaty, rancid	Octanoic acid	124-07-2	1990		–	2070
C	Rancid, sweaty	Decanoic acid	334-48-5	370		–	2277
C	Rancid, sweaty	9-Decenoic acid	14436-32-9	150		–	2335
C	Floral rose	Phenylacetic acid	103-82-2	2080		–	2556

Independent contributions to overall aroma; "+" indicate the odorant that was statistically significant in the triangle difference test with p value ≤ 0.05. The differential threshold values in beer were determined in the current study using the reference beer as base beer contributions of each of 38 odorants in detail, the independent contribution of each odorant to overall beer aroma was examined by a triangle difference test, using the reference Pilsner-type beer as the base solution. Although 9 odorants were present at concentrations above their threshold levels, the results of the triangular tests (in Table 5.1) showed that only 6 odorants with higher OAVs – isoamyl acetate, 3-methyl-1-butanol, 2-phenylethyl alcohol, γ-nonalactone, furaneol, and sotolon – in group A were significant and have independent contribution to overall beer aroma. Thus the investigations suggested the possibility that the odorants at sub-threshold levels were insignificant in the triangle test and do not contribute independently to the overall characteristics.

Of the prescreened 38 odorants, 25 odorants with higher OAVs (Table 5.1) ranging from 3.58 to 0.35, which were supposed to construct the base framework structure of the beer aroma, were then selected and grouped as A in Table 5.1. An aroma recombinant containing the 25 components at the same concentrations as they occurred in the beer was established in a water containing 0.20 MPa of CO_2 and 40,900 mg/L of ethanol, with pH adjusted to 4.10.

As depicted in Fig. 5.1a, it could be observed that the recombinant using 25 odorants of group A remarkably lacked the malty/cereal aroma (with a score of −2.14), estery flavor (−1.57), and thickness (−2.00) and contained outstanding

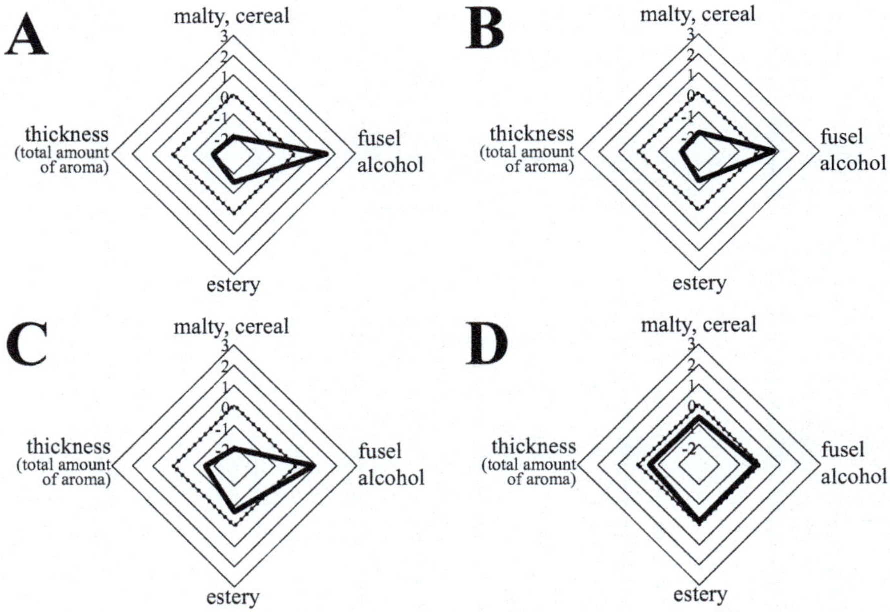

Fig. 5.1 Aroma profile analyses of the aroma recombinants comprising 25 odorants of groups A (**a**), 49 odorants of A + B (**b**), 52 odorants of A + C (**c**), and 76 odorants of A + B + C (**d**) (Kishimoto et al. 2018)
The area delineated with the dotted border indicates the aroma profile of the reference Pilsner-type beer

intensity of unbalanced fusel alcohol notes (+1.57) mainly derived from the odorants 3-methyl-1-butanol, methionol, and 2-phenylethyl alcohol. This indicated that the recombinant lacked the characteristics that balance with the high intensity of fusel alcohol notes.

The similarity was also evaluated by comparison with the reference beer, whose score was set at 100%. The aroma quality of the group A recombinant with unbalanced aroma characteristics differed substantially from that of the reference beer and as the result of a detailed discussion among the panelists, the aroma similarity of the group A recombinant with such insufficient characteristics was set at 30%. In the following experiments, the similarities of the recombinants were evaluated by comparison with the reference Pilsner-type beer and the group A recombinant as the markers.

5.2.5 *Investigation of the Malty Aroma Fraction*

To compensate for the characteristics that were lacking remarkably in the group A recombinant, the aroma components contributing to the malty/cereal flavor were investigated. The aroma extract of reference beer (extract Q) was fractionated using aminopropyl group-modified silica gel column, with CH_2Cl_2/methanol as the eluent. Then the malty/cereal flavor of each fraction was examined.

The aminopropyl group-modified silica gel column was employed in the fractionation of the current study to prevent the tailing on silica gel. The existence of basic compounds such as nitrogen-containing compounds and their contributions to malty/cereal flavor were often reported (Harding et al. 1977). Basic compounds tail on silica gel because they interact with the slightly acidic silica, and it was impossible to yield the precise fractions with clear boundaries when the standard silica gel was used for the filler of the column.

As the result of fractionation of extract Q, concentrated fractions of Q-No.4 and 6, eluted with CH_2Cl_2/methanol with the ratios of 40:1 and 20:1 by volume, respectively, elicited a malty or roasty aroma. The concentrated fraction Q-No.4 elicited characteristics of cereals, nuts, almonds, and weak earthy scents, which seemed to be derived from the odor qualities of 2-acetyl-1-pyrroline, 2,3,5-trimethylpyrazine, 2-ethyl-3,5(6)-dimethylpyrazine, benzaldehyde, γ-hexalactone, γ-octalactone, and vanillin. Fraction Q-No.6 elicited caramel-like, burnt sugar, and sweet aromas that appeared to be derived from the odor qualities of benzyl mercaptan, cyclotene, maltol, γ-nonalactone, furaneol, homofuraneol, sotolon, 4-vinylguaiacol, and maple furanone.

The odorants listed in Table 5.1 were detected in the GC-O analyses of concentrated fractions of Q-No.4 and 6. Of these odorants, 44 odorants that appeared to contribute to the characteristics of Q-No.4 and 6 were selected based on their Charm values and aroma qualities and then identified. Among the 44 odorants, 20 were already included in group A and the remaining 24 odorants were grouped as B in Table 5.1. The 24 odorants of group B, which were expected to contribute to the malty/cereal flavor, were added to the group A recombinant to examine their contributions.

As depicted in Fig. 5.1b, the recombinant containing 49 components of groups A and B still remarkably lacked a malty/cereal character (with a score of −2.04), estery character (−1.50), and thickness (−2.14), with a higher intensity of fusel alcohol characteristics (+0.75) and lower similarity evaluated as 29.4% (Fig. 5.2). That is, the deficient aroma qualities and the similarity of the group A recombinant were not improved by the addition of the group B components.

The contributions of odorants in group B were examined using the triangle difference test described previously. Each of the 24 odorants in group B was insignificant in the triangular test (Table 5.1) and every component was revealed to have no independent contribution to the overall beer aroma.

5.2.6 Other Odorants with No Independent Contributions to Overall Aroma Characteristics

Of all the 76 odorants that were identified after the GC-O analyses of extract P and the entire fractions of extract Q, 49 odorants were already included in the components of groups A or B. The remaining 27 odorants were classified as group C (Table 5.1), and their contributions to overall beer aroma were examined using the triangle difference test described earlier. Each of the 27 odorants in group C was

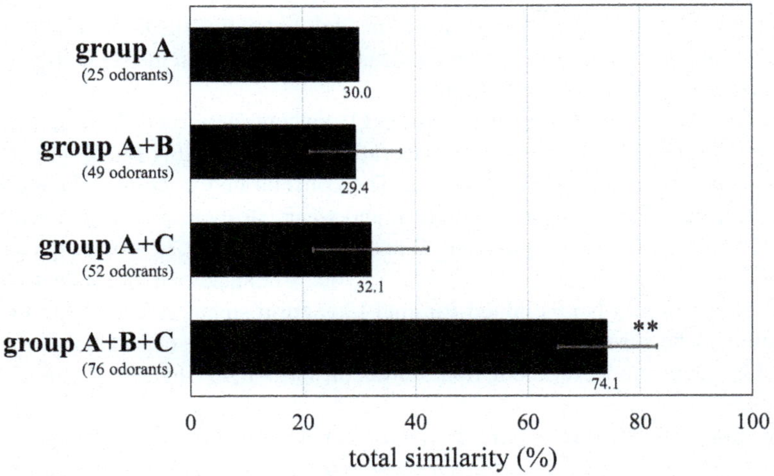

Fig. 5.2 Total similarity of aroma recombinants comprising groups A, A + B, A + C, and A + B + C (Kishimoto et al. 2018)
The data indicates the averages and standard deviations of the similarity score among the panelists. The asterisks (**) indicate that the difference is statistically significant with p value ≤0.05

insignificant in the triangular test (Table 5.1) and every component was revealed to have no independent contribution to overall beer aroma.

The group C components with no independent contributions to the overall aroma characteristics were added to establish the recombinant containing all 76 components of group A, B, and C. As the result of the sensory test depicted in Fig. 5.1d, the characteristics of aroma recombinant containing all 76 components improved dramatically, with the score of malty/cereal character increasing to −0.64, the fusel alcohol characteristics to −0.21, the estery characteristics to −0.14, and the thickness to −0.57. The similarity of the recombinant with well-balanced characteristics improved dramatically to 74.1% (Fig. 5.2).

Moreover, the 27 components of group C were added to the group A recombinant to establish the recombinant with 52 components of groups A and C. The sensory test of the established recombinant containing 52 components of groups A and C exhibited a lower similarity of 32.1% (Fig. 5.2). That is, the deficient characteristics of the group A recombinant were not improved by the addition of only group C components, as depicted in Fig. 5.1c.

5.3 Conclusion

Not only 27 odorants in group C, but the addition of 51 odorants in group B and C were required to improve the characteristics and similarity of the recombinant. The investigation indicated that the synergistic contributions of multiple odorants, with no independent contribution to overall aroma characteristics (including subthreshold components), balanced with the outstanding fusel alcohol notes and constructed the well-balanced characteristics.

References

Acree TE (1993) Bioassays in flavor. In: Acree TE, Teranishi R (eds) Flavor science. Sensible principles and techniques, ACS professional reference book. American Chemical Society, Washington, DC, pp 1–22

Acree TE, Barnard J (1996) A method for the measurement of odor. In: A tutorial for charm analysis. DATU, New York, pp 7–8

Arrieta-Baez D, Dorantes-Álvarez L, Martinez-Torres R, Zepeda-Vallejo G, Jaramillo-Flores ME, Ortiz-Moreno AG, Aparicio-Ozores R (2012) Effect of thermal sterilization on ferulic, coumaric and cinnamic acids: dimerization and antioxidant activity. J Sci Food Agric 92:2715–2720

Blank I (2002) Gas chromatography-olfactometry in food aroma analysis. Food Sci Technol 115:297–331

Cabrita MJB, Garcia R, Martins N, Gomes Da Silva M, Freitas AMC (2012) Gas chromatography in analysis of compounds released from wood into wine. In: Advanced gas chromatography – progress in agricultural, biomedical and industrial applications, pp 185–298. https://doi.org/10.5772/2518

European Brewery Convention (2010) Bitterness of beer (IM), EBC methods of analysis, section 9 beer method 8. In: Analytica-EBC. Fachverlag Hans Carl, Nürnberg

Fiddler W, Parker WE, Wasserman AE, Doerr RC (1967) Thermal decomposition of ferulic acid. J Agric Food Chem 15:757–761

Grosch W (2001) Evaluation of the key odorants of foods by dilution experiments, aroma models and omission. Chem Senses 26:533–545

Harding RJ, Nursten HE, Wren JJ (1977) Basic compounds contributing to beer flavour. J Sci Food Agric 28:225–232

Kaneda H, Kobayashi N, Watari J, Shinotsuka K, Takashio M (2002) A new taste sensor for evaluation of beer body and smoothness using a lipid-coated quartz crystal microbalance. J Am Soc Brew Chem 60:71–76

Kishimoto T, Wanikawa A, Kagami N, Kawatsura K (2005) Analysis of hop-derived terpenoids in beer and evaluation of their behavior using the stir bar-sorptive extraction method with GC-MS. J Agric Food Chem 53:4701–4707

Kishimoto T, Wanikawa A, Kono K, Shibata K (2006) Comparison of the odor-active compounds in unhopped beer and beers hopped with different hop varieties. J Agric Food Chem 54:8855–8861

Kishimoto T, Wanikawa A, Kono K, Aoki K (2007) Odorants comprising hop aroma of beer: hop-derived odorants increased in the beer hopped with aged hops. In: Proceedings of the 31st European Brewery Convention Congress. Getränke-Fachverlag Hans Carl, Nürnberg, pp 226–235

Kishimoto T, Kobayashi M, Yako N, Iida A, Wanikawa A (2008) Comparison of 4-mercapto-4-methylpentan-2-one content in hop cultivars from different growing regions. J Agric Food Chem 56:1051–1057

Kishimoto T, Noba S, Yako N, Kobayashi M, Watanabe T (2018) Simulation of Pilsner-type beer aroma using 76 odor-active compounds. J Biosci Bioeng 126:330–338

Kurobayashi Y, Katsumi Y, Fujita A, Morimitsu Y, Kubota K (2008) Flavor enhancement of chicken broth from boiled celery constituents. J Agric Food Chem 56:512–516

Mayol AR, Acree TE (2001) Advances in gas chromatography-olfactometry. ACS Symp Ser 782:1–10

Miyazawa T, Gallagher M, Preti G, Wise PM (2008) Synergistic mixture interactions in detection of perithreshold odors by humans. Chem Senses 33:363–369

Schieberle P (1995) Recent developments in methods for analysis of flavor compounds and their precursors. In: Goankar A (ed) Characterization of Food: emerging methods. Elsevier, Amsterdam, pp 403–431

Sensory analysis (1992) Methods of analysis of the American Society of Brewing Chemists. American Society of the Brewing Chemists, St Paul

Chapter 6
Koku Attribute-Enhancing Odor Compounds

Yoshiko Kurobayashi, Satoshi Fujiwara, Tomona Matsumoto, and Akira Nakanishi

Abstract *Koku* is a popular Japanese term describing improved "complexity," "continuity," and "mouthfulness" as attributes of palatability when experiencing flavor. While *koku* has long been thought to derive from taste substances and textural sensations, certain kinds of odor compounds have been reported as "*koku* attribute" enhancers. In this chapter we will present three cases to illustrate the possibility that a single volatile compound in a given food can produce an odor sensation to enhance the "*koku* attribute." The cases introduce the effects of the three types of volatile compounds: phthalide compound in celery, (4Z,7Z)-trideca-4,7-dienal in dried bonito, and rotundone in several fruits.

Keywords Odor compound · Phthalide · (4Z,7Z)-Trideca-4,7-dienal · Rotundone · QDA (quantitative descriptive analysis) · NIRS (near-infrared spectroscopy)

6.1 Introduction

Koku is a Japanese term popularly used to describe the sense of palatability derived from improvements in "complexity," "continuity (lingeringness)," and "mouthfulness" when experiencing flavor (Nishimura and Egusa 2015). While the term is applied to a wide range of foods and beverages such as well-cooked soup to coffee, tea, cocoa, dairy products, alcohol beverages, and fruit juice, the nuances or attributes of *koku* differ markedly from one food to another. This variability has given rise to a concept of "*koku* attributes" that has spread from Japan to many other parts of the world, especially to western culinary circles. Yet the term *koku* and concept and definition of "*koku* attributes" remain controversial and somewhat amorphous.

While "*koku*" has long been thought to derive from taste substances and textural sensations, certain kinds of odor compounds have been reported as "*koku* attribute" enhancers. Kawada and Katsumata reported that pyrazines contribute to a *koku* sen-

Y. Kurobayashi (✉) · S. Fujiwara · T. Matsumoto · A. Nakanishi
T. Hasegawa Co. Ltd. R & D Center, Nakahara-ku, Kawasaki-shi,
Kanagawa Prefecture, Japan
e-mail: yoshiko_kurobayashi@t-hasegawa.co.jp

T. Nishimura, M. Kuroda (eds.), *Koku in Food Science and Physiology*,
https://doi.org/10.1007/978-981-13-8453-0_6

sation in well-simmered stew or long-fermented soy paste, along with Maillard peptides and amino acids (Kawada and Saitou 1998; Katsumata 2014). Hayase et al. found that the addition of 2-acetylfuran, 2-ethylhexanol, or 1-octen-3-ol at levels far below their thresholds markedly enhanced the "*koku*" taste of seasoning soy sauce (Hayase et al. 2013). Nishimura et al. demonstrated that heat-treated onion aroma bound to stigmasterol and β-sitosterol caused a lingering flavor in chicken consommé (Nishimura et al. 2016).

Our own research on the flavor of chicken broth have reported the flavor-enhancing effects of phthalides, the volatile compounds in celery, based on a set of "*koku* attributes" sensed in chicken soup or bouillon ("thick," "impactful," "mild," "lasting," "satisfying," "complex," "clarified," "sweet," and "umami"), and has demonstrated the importance of odor compounds for those attributes (Kurobayashi et al. 2008). In this chapter we present more examples to illustrate how the odor sensation elicited by a single volatile compound in food may enhance the "*koku* attributes."

6.2 Phthalides in Celery Volatiles

In practical cuisine, animal materials such as meat and bone are simmered in water together with plant materials such as aromatic vegetables and spices over a period of hours to produce a stock or bouillon with a rich, complex flavor. Celery is usually regarded as an essential ingredient to reduce off-flavor and increase the desirable complex flavor in stocks of this sort. Among the volatile constituents of celery, phthalide skeleton structure compounds such as 3-*n*-butylphthalide, sedanenolide, and *trans-/cis*-sedanolide have been identified as major contributors to the characteristic spicy note of celery (Fig. 6.1). We conducted experiments to verify the hypothesis that these compounds also contribute to the flavor-enhancing effect of celery on simmered dishes (Kurobayashi et al. 2008).

6.2.1 Experimental

Several constituents related to celery were evaluated by preparing sample solutions of the constituents mixed into a chicken broth preliminarily prepared from chicken meat and bone alone. A panel of 12 female evaluators well trained in sensory

Fig. 6.1 Characteristic odor compounds in celery

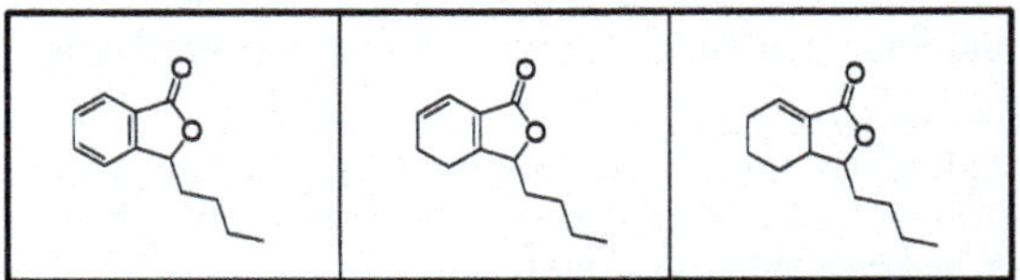

3-*n*-butylphthalide	sedanenolide	*trans-/cis*-sedanolide

evaluation assessed the differences in flavor between the sample solutions and chicken broth only as control by means of quantitative descriptive analysis (QDA). The evaluators were asked to taste the samples buccally for several seconds and then to rate the intensity of the following 10 attributes characterizing the "*koku*" of chicken broth on a linear scale relative to a control value of zero "thick," "impactful," "mild," "lasting," "satisfying," "complex," "refined," "clarified," "umami," and "sweet." Three kinds of evaluations were performed by the above method.

The first evaluation focused on the effects of volatile and nonvolatile constituents of celery on chicken broth flavor. The volatile fraction was a distillate obtained from a steam distillation of celery and the nonvolatile fraction was a residual fraction of the distillation. While the former possessed a strong and characteristic celery-like odor with no taste, the latter possessed a strong, sweet, umami taste with a weak, sweet, boiled vegetable-like odor. The test solutions were prepared by adding the volatile and nonvolatile fractions to the chicken broth at levels estimated to approximate those present in a practical soup preparation without showing a distinct celery-like odor.

The second evaluation focused on the effects of the three phthalides sedanenolide, 3-*n*-butylphthalide, and sedanolide added alone to chicken broth at levels of 0.2 ppm, 0.7 ppm, and 0.2 ppm, respectively.

The third examination focused on the gustatory sensations of phthalide compounds. A triangle test was conducted using a duplicated control and odd sample containing one phthalide compound at the same level described above. The panel was asked to taste the three samples buccally and to identify the odd samples in two sessions, one with noseclips to prevent the flow of odor to the olfactory epithelium and one without noseclips.

6.2.2 Results

6.2.2.1 Effects of Volatile and Nonvolatile Fractions

Figure 6.2 shows the results of the evaluation of chicken broth flavors with the volatile and nonvolatile fractions of celery in comparison to the control. All of the scores of both test solutions were higher than zero, and most scores were higher in the broth mixed with the volatile fraction. Hence the volatile constituents enhanced a deeper chicken broth flavor than the nonvolatile constituents. "Umami" and "sweet," attributes generally recognized as taste sensations, were enhanced by the volatile fraction of celery to the same degree that they were with the nonvolatile fraction. These results indicate that the degree to which the complex flavor and taste of chicken broth are enhanced depends more on the odor components than on the taste components.

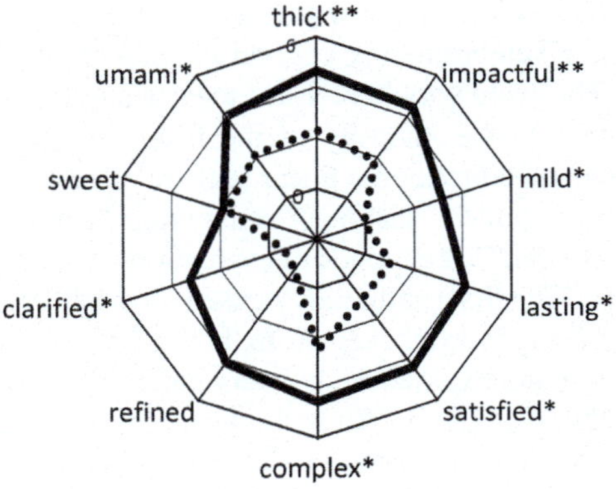

Fig. 6.2 Effects of volatile and nonvolatile fractions of celery on chicken broth flavor
▬ Chicken broth added volatile fraction of celery
•••• Chicken broth added nonvolatile fraction of celery
$*p < 0.05$, $**p < 0.01$

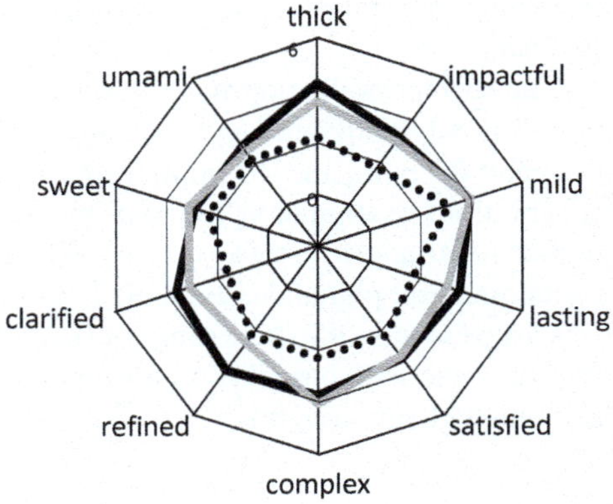

Fig. 6.3 Effects of three phthalides on chicken broth flavor
▬ Chicken broth added sedanenolide
•••• Chicken broth added 3-*n*-butylphthalide
▬ Chicken broth added sedanolide

6.2.2.2 Effects of the Three Phthalides

The next evaluation focused on the effects of 3-*n*-butylphthalide, sedanenolide, and sedanolide. While the test solution showed no distinctive characteristic celery-like odor at the levels conducted in this evaluation, all attributes scored higher than the control when the test samples were buccally tasted (see Fig. 6.3). While no differences were observed in nearly any of the attributes of the three samples, 3-*n*-butylphthalide tended to score lower for the attributes "thick" and "lasting" than sedanenolide and sedanolide, phthalides that confer the same effects as the volatile fraction of celery. These results indicate that the phthalides are responsible for the effect of the celery volatiles in enhancing the complex flavor and taste of chicken broth.

Table 6.1 Number of correct responses for detection of phthalides in chicken broth with and without noseclip in the triangle test

	Number of correct answers	
	Without noseclip	With noseclip
3-*n*-Butylphthalide	6	1
Sedanenolide	9***	1
Sedanolide	9***	1

$n = 10$
*** $p < 0.001$

6.2.2.3 Effects of the Phthalides on Gustatory Sensation

Next, we ascertained which sensation was involved in these effects via gustatory or olfactory mechanisms. Table 6.1 shows the number of correct responses in detecting the test samples with and without the noseclips attached. The evaluators found no differences in flavors when wearing the noseclips but did without the clips. This result shows that the phthalides have only weak gustatory property; hence, we can assume that the olfactory sensation elicited by the phthalides enhanced the complex *koku* flavor of chicken broth.

6.3 (4Z,7Z)-Trideca-4,7-Dienal in Broth of Dried Bonito

Dried bonito is widely used as a traditional food ingredient in Japanese cuisine. It is made from bonito through a lengthy process of boiling, drying, and smoking that finally confers a very solid character and substantial umami and flavor constituents. Dried bonito is typically used to prepare a broth: shaved dried bonito is put into boiled water and left for several minutes to provide the rich taste and complex aroma, or the *koku* attribute. In this study we assumed that the rich flavor may derive from odor compounds as well as taste substances such as amino acids, fatty acids, nucleic acids, and peptides.

The many studies so far performed on the flavor components of dried bonito have identified more than 400 compounds. Besides these compounds, our group found (4Z,7Z)-trideca-4,7-dienal (hereafter TDD) at a level of less than 10 ppb as a potent odor compound in dried bonito (Saito et al. 2014). Because of its strong odor with a discernible woody, cardboard-like note, we decided to focus on this compound in our present experiments (Fujiwara et al. 2015).

6.3.1 *Experiment*

The effects of TDD on dried bonito broth were evaluated by measuring the cerebral blood flow as a hemodynamic response by near-infrared spectroscopy (NIRS), along with the sensory evaluation, as blood flow changes measured on the temple region are reported to reflect the secretion of saliva, a putative indicator of appetitive motivation (Sato et al. 2011).

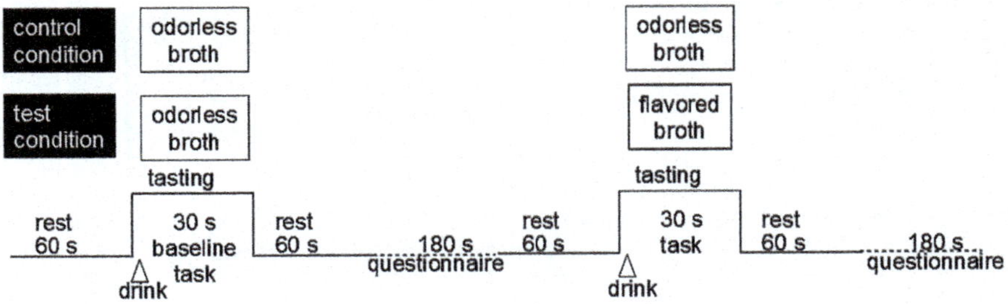

Fig. 6.4 Experimental protocol of measurement of hemodynamic responses and sensorial evaluation

The measurements were conducted on three samples. Sample 1, a model taste solution consisting of amino acids, nucleic acids, and NaCl, mimicked odorless dried bonito broth. Samples 2 and 3 were prepared by combining the sample 1 solution with a reconstituted dried bonito odor prepared with standard chemicals according to analytical data to mimic a flavored dried bonito broth, without and with 5 ppb of TDD, respectively. The evaluations were done by ten participants.

The salivary hemodynamic responses were measured with an ETG-400 Optical Topography System (Hitachi Medical Co., Tokyo, Japan) using a 3x11 optode set with a total of 52 channels. Figure 6.4 shows the measurement procedure. Next, the panel performed a QDA as a sensory evaluation. The participants were asked to rate the intensities of six attributes used to described the character of dried bonito broth: "palatability," "aroma intensity," "umami intensity," "thickness," "consonance," and "complexity."

6.3.2 Results

Figure 6.5a shows the relative intensities of the hemodynamic responses of sample 2, the flavored model solution without TDD, to sample 1, the control. Contrary to expectations, an increase in the hemodynamic responses was observed in only five of the ten participants (a-1), while no change was observed in the hemodynamic responses of the other 5 (a-2). In the comparison with sample 3, the flavored model solution with TDD, however, a significant increase in hemodynamic responses was observed in both the latter participants (b-2) and former participants (b-1) (see Fig. 6.5b). That is, the salivary responses of all participants were enhanced when they tasted the dried bonito-like flavored broth with TDD.

The results of the QDA for the former five participants and latter five participants are shown separately in two plots in Fig. 6.6. The former five participants assigned higher scores to the attributes of "palatability," "aroma intensity," "thickness," and "complexity" in the evaluation of sample 2. The latter participants, meanwhile, assigned much lower scores to the attributes "consonance" and "palatability." The former participants appeared to sense congruence in the dried bonito-like odor and taste in sample 2, the flavored model solution without TDD, and thus to sense

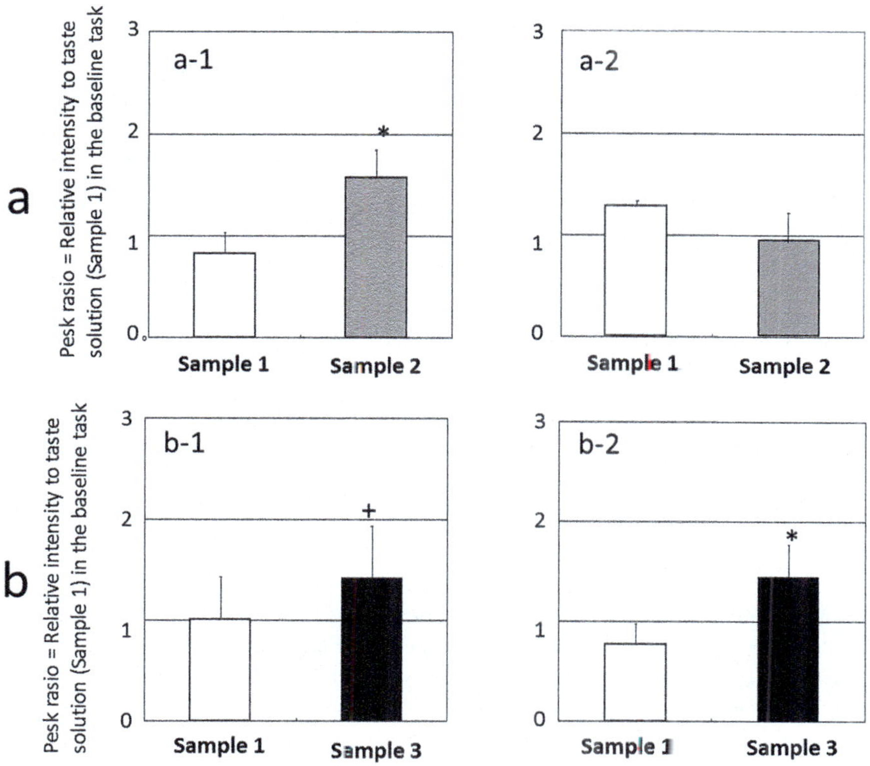

Fig. 6.5 Relative intensity or hemodynamic responses of sample 1–3 to control

Model taste solution (Sample 1)

Model taste solution added dried bonito-like flavor wtihout TDD (Sample 2)

Model taste solution added dried bonito-like flavor with TDD (Sample 3)

Relative scores in **a**-1 and **b**-1 are average scores of the same five participants giving high response in sample 2

Relative scores in **a**-2 and **b**-2 are average scores of the same five participants giving no change in sample 2

$+p < 0.1$, $*p < 0.05$

enhanced “complexity” and “thickness,” the important factors of *koku*. As a consequence, the former participants rated the sample higher for overall palatability and demonstrated higher salivary hemodynamic responses, or appetitive motivation. The latter participants, meanwhile, appeared to feel no congruency in the taste and odor in sample 2 and might have felt a strange or imbalanced quality. As a consequence, they felt much less “palatability” and exhibited no hemodynamic responses.

Figure 6.6 also shows the results of the evaluations of sample 3, the flavored model solution with added TDD. While the former participants gave almost the same scores as they did for sample 2, the latter gave higher ratings, especially for

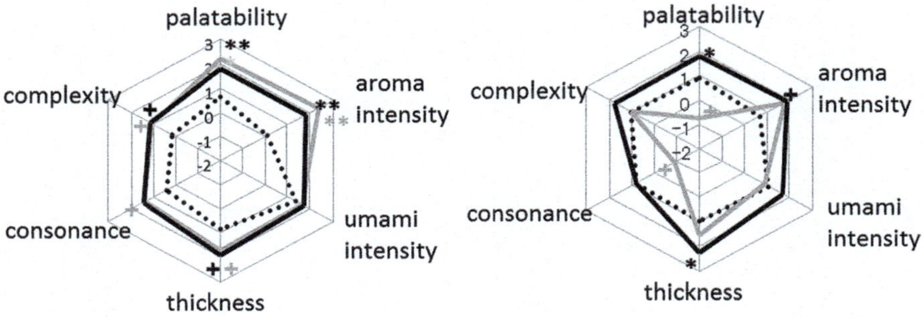

Fig. 6.6 Sensory evaluation of odoreless and dried bonito-flavored model solution
.... Model taste solution (sample 1)
Model taste solution added dried bonito-like flavor wtihout TDD (sample 2)
Model taste solution added dried bonito-like flavor with TDD (sample 3)
The scores in the left side (*1*) were the average scores of the same five participants as in Fig. 6.5-**a**-1 and **b**-1
The scores in the right side (*2*) were the average scores of the same five participants as in Fig. 6.5-**a**-2 and **b**-2
$+p < 0.1$; $*p < 0.05$; $**p < 0.01$

the attributes "consonance" and "palatability." These results indicate that the addition of TDD may have led to perceptions of congruency of taste and odor, to a flavor closer to the desirable real dried bonito broth, and, as a consequence, to a higher score for palatability and higher hemodynamic responses.

These results convinced us that TDD contributes powerfully to the overall flavor of dried bonito broth, especially in the attribute of "*koku*."

6.4 Rotundone in Several Fruits

Most fruits are loved the world over for pleasant flavors composed of odor compounds of various kinds (fruits, for our purposes, mean edible fruits). Their light and refreshing flavors are sometimes recognized as having less "*koku*" (Nishimura and Egusa 2016). Our group, however, has empirically observed that natural fruit has a deeper flavor than artificial fruit juice. Here, therefore, we tried to reconsider fruit flavors in terms of "*koku*."

Though numerous studies have revealed the characteristic flavor compounds of various kinds of fruit, the flavor of natural fruit has been difficult to reconstitute. Our attempts to analyze fruit flavors by gas chromatography/olfactometry (GC/O) have commonly detected a particular compound with a woody note in several kinds of fruit. After a series of examinations, we identified this woody-note compound as 5-isopropenyl-3,8-dimethyl-3,4,5,6,7,8-hexahydro-1(2*H*)-azulenone (Nakanishi et al. 2017).

This compound had already been identified and given the name rotundone. Rotundone was first found in *Cyperus rotundus*, a natural herbal remedy in China, in 1967 (Kapadia et al. 1967), and was later identified in agarwood (Ishihara et al. 1991). More recently it has been identified as a peppery flavor contributor in Shiraz grape and Shiraz wine and as a spicy note contributor in black pepper, white pepper, rosemary, and so on. The odor threshold has been quantified at 8 ppt (Wood et al. 2008). No one before us, however, has identified rotundone in fruit flavor. We assumed that rotundone contributes not only to peppery flavor, but also to other attributes in the flavors of many kinds of food.

6.4.1 Experimental

A sensory evaluation was conducted to assess the effects of rotundone. Four kinds of fruit were evaluated: grapefruit, orange, mango, and apple. Test samples were prepared with taste solution, with sugar and citric acid, and with four kinds of reconstituted odor prepared based on quantitative data reported in the corresponding references (Buettner and Schieberle 2001a, b; Steinhaus et al. 2006; Boonbumrung et al. 2001). The panel consisted of six male and four female evaluators, all with thorough training in sensory evaluation. The attributes for each fruit were terms collaboratively selected by the panel as descriptors of the respective fruit flavor characters: for grapefruit, "sweet," "sour," "bitter," "fresh," "juicy," "complex," and "discordant"; for orange, "sweet," "sour," "floral," "fresh," "juicy," "peely," "complex," and "discordant"; for apple, "sweet," "sour," "green," "fruity," "pulpy," "complex," and "discordant"; for mango, "sweet," "sulfurous," "green," "fruity," "metallic," "matured," "complex," and "discordant."

In each evaluation, the panel was asked to taste test samples 1 and 2, the samples without and with 5 ppt of rotundone, and to rate the intensity of each attribute.

6.4.2 Results

Figure 6.7 shows average scores for the attributes in the evaluations of grapefruit, orange, apple, and mango. While the attributes themselves varied in the four evaluations, a significant decrease in the score for "discordant" and a significant increase in "complex" were commonly observed in sample 2. The evaluators also sensed that other desirable attributes were enhanced in sample 2 with rotundone, such as "fresh," "peely," and "sour" in grapefruit;" peely," "flowery," "fresh," and "juicy" in orange; "matured" in mango; and "pulpy" in apple. The attribute "discordant" in these valuations was defined as "unpleasant taste" or "rough flavor," while "complex" was defined as "deep" and "thick." Thus, a decrease of "discordant" and increase of "complex" and other desirable attributes seemed to lead to an enhanced "*koku*" sensation in the fruit flavors.

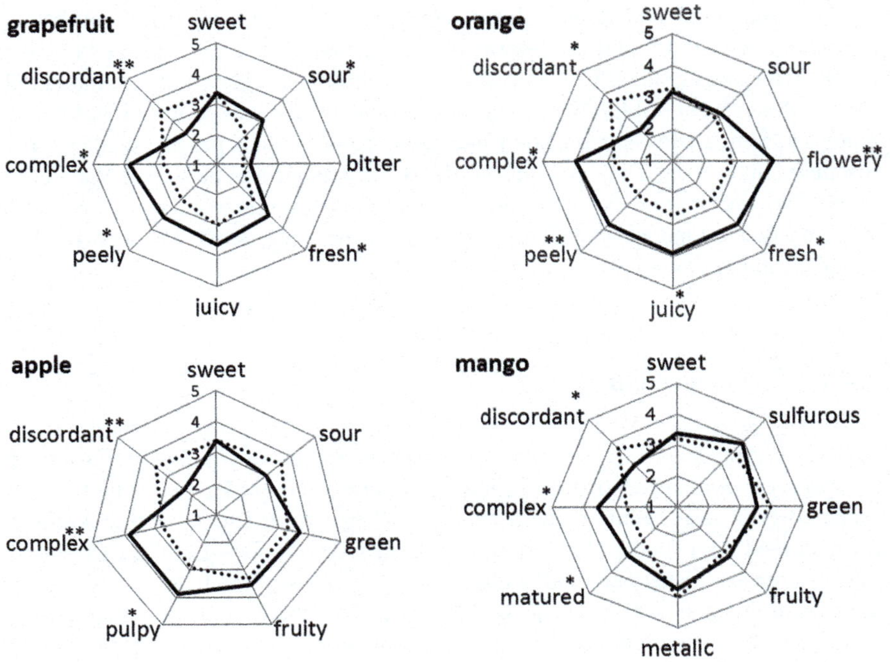

Fig. 6.7 Effects of rotundone on fruit flavors
▪▪▪▪ Sample 1 taste solution + aroma reconstitute
▬ Sample 2 taste solution + aroma reconstitute with rotundone
$*p < 0.05$; $**p < 0.01$

From these results, we find that the sense of "*koku*" can be enhanced by the addition of the odor compound rotundone even in fruits, a type of food generally thought to have less *koku* attribute.

6.5 Conclusion

These examinations demonstrated that three kinds of odor compounds, namely, the phthalides, TDD, and rotundone, are independently effective in modifying and enhancing the flavors of foods. Their modifying and enhancing effects are observed even when the compounds are added at subthreshold levels. The changes they bring about are not changes of a specific character such as the spicy note of phthalides, the woody or cardboard-like note of TDD, or the woody note of rotundone. Rather, they are changes of a complex, well-balanced, desirable character, or what we describe as the "*koku* attribute."

In conclusion, we have demonstrated the prospect of applying a single odor compound to enhance the "*koku* attribute" of foods or dishes. Further, we propose that fruits, which are normally thought to have less *koku*, in fact have a "*koku* attribute" that can be enhanced by odor compounds.

References

Boonbumrung S, Tamura H, Moookdasanit J, Nakamoto H, Ishihara M, Yoshizawa T, Varanyanond W (2001) Characteristic aroma components of the volatile of yellow keaw mango fruits determined by limited odor unit method. Food Sci Technol Res 7:200–206

Buettner A, Schieberle P (2001a) Evaluation of key aroma compounds in hand-squeezed grapefruit juice (*Citrus paradise* Macfayden) by quantitation and flavor reconstitution experiments. J Agric Food Chem 49:1358–1363

Buettner A, Schieberle P (2001b) Evaluation of aroma difference between hand-squeezed juice from valencia late and navel oranges by quantitation of key odorants and flavor reconstitution experiments. J Agric Food Chem 49:2387–2394

Fujiwara S, Nakamura A, Mori K, Watanabe H, Kurobayashi Y, Saito T, Fushiki T (2015) A study on the potent odour compound of dried bonito. In: Taylor AJ, Mottram DS (eds) Flavour Science Proceedings of the XIV Weurman Flavour Research Symposium. Context Products Ltd, Leicestershire, pp 451–454

Hayase F, Takahagi Y, Watanabe H (2013) Analysis of cooked flavor and odorants contributing to the koku taste of seasoning soy sauce. Nippon Shokuhin Kagaku Kogaku Kaishi 60:59–71

Ishihara M, Tsuneya T, Uneyama K (1991) Guaiene sesquiterpene from agarwood. Phytochemistry 30:3343–3347

Kapadia VH, Naik VG, Wadia MS, Dev S (1967) Sesquiterpenoids from the essential oil of *Cyperus rotundus*. Tetrahedron Lett 8:4661–4667

Katsumata T (2014) Shokuhin no koku to kokumi choumiryou. Gekkan Food Chem 8:40–44

Kawada S, Saitou C (1998) Kokumi choumiryou no tokusei to riyou. Food Sci 40:87–91

Kurobayashi Y, Katsumi Y, Fujita A, Morimitsu Y, Kubota K (2008) Flavor enhancement of chicken broth from boiled celery constituents. J Agric Food Chem 56:512–516

Nakanishi A, Fukushima Y, Miyazawa N, Yoshikawa K, Masuda Y, Kurobayashi Y (2017) Identification of rotundone as a potent odor-active compound of several kinds of fruits. J Agric Food Chem 65:4464–4471

Nishimura T, Egusa A (2016) "Koku" involved in food palatability an overview of pioneering work and outstanding questions. Kagaku to Seibutsu 54:102–108

Nishimura T, Egusa A, Nagano A, Odahara T, Sugise T, Mizoguchi N, Nosho Y (2016) Phytosterols in onion contribute to a sensation of lingering of aroma, a koku attribute. Food Chem 192:724–728

Saito T, Shiibashi H, Myoga H, Harabuchi K, Masuda Y, Kurobayashi Y, Nammoku T, Yamazaki H, Nakamura M, Fushiki T (2014) Aroma compounds contributing to dried Bonito. Nippon Shokuhin Kagaku Kogaku Kaishi 61:519–527

Sato H, Obata A, Moda I, Ozaki K, Yasuhara T, Yamamoto Y, Kiguchi M, Maki A, Kubota K, Koizumi H (2011) Application of near-infrared spectroscopy to measurement of hemodynamic signals accompanying stimulated saliva secretion. J Biomed Opt 16(4):047002

Steinhaus M, Bogen J, Schieberle P (2006) Key aroma compounds in apple juice – changes during juice concentration. In: WLP B, Peterson MA (eds) Flavour science recent advances and trends. Elsevier, Washington, DC, pp 189–192

Wood C, Siebert TE, Parker M, Capone DL, Elsey GM, Pollnitz AP, Eggers M, Meier M, Voessing T, Widder S, Krammer G, Seftor MA, Herderich MJ (2008) From wine to pepper: rotundone, an obscure sesquiterpene, is a potent spicy aroma compound. J Agric Food Chem 56:3738–3744

Chapter 7
Effect of a *Kokumi* Peptide, γ-Glutamyl-Valyl-Glycine, on the Sensory Characteristics of Foods

Motonaka Kuroda and Naohiro Miyamura

Abstract Certain foods are known to have flavor characteristics such as continuity and mouthfulness which cannot be explained by the five basic tastes alone. It has been demonstrated that these sensations are evoked by the addition of *kokumi* substances, flavor modifiers with no taste of their own, which are considered to be one of "*koku*"-imparting factors. Recent studies have demonstrated that *kokumi* substances such as glutathione are perceived via the calcium-sensing receptor (CaSR). Screening with CaSR assay and sensory evaluations have shown that γ-glutamyl-valyl-glycine (γ-Glu-Val-Gly) is a potent *kokumi* peptide. In the present study, the sensory effects of γ-Glu-Val-Gly on chicken consommé, reduced-fat peanut butter, and reduced-fat custard cream were investigated by descriptive analysis. Chicken consommé containing γ-Glu-Val-Gly had significantly stronger umami, mouthfulness, and mouth-coating characteristics than those of the control. In addition, γ-Glu-Val-Gly significantly increased the oiliness, thick flavor, and aftertaste of reduced-fat peanut butter, significantly increased the thickness of taste of reduced-fat custard cream, and significantly increased the aftertaste of reduced-fat French dressing. These data indicated that a *kokumi* peptide, γ-Glu-Val-Gly, can enhance umami, mouthfulness, and mouth-coating of chicken consommé; enhance oiliness, thickness of taste, and aftertaste of reduced-fat peanut butter; enhance thickness of taste of reduced-fat custard cream; and enhance aftertaste of reduced-fat French dressing. As a *kokumi* substance, γ-Glu-Val-Gly has no taste of its own; it can therefore be used to improve the flavor of both savory and sweet foods.

Keywords *Kokumi* substance · γ-Glutamyl-valyl-glycine · γ-Glu-Val-Gly · Glutathione · GSH · Chicken consommé · Peanut butter · Custard cream · French dressing · Descriptive analysis

---

M. Kuroda (✉) N. Miyamura
Institute of Food Sciences & Technologies, Ajinomoto,
Suzuki-cho, Kawasaki, Kanagawa, Japan
e-mail: motonaka_kuroda@ajinomoto.com

T. Nishimura, M. Kuroda (eds.), *Koku in Food Science and Physiology*,
https://doi.org/10.1007/978-981-13-8453-0_7

Abbreviations

γ-Glu-Val-Gly	γ-Glutamyl-valyl-glycine
GSH	Glutathione
CaSR	Calcium-sensing receptor
FEMA	Flavor and extract manufacturers association
IMP	Inosine monophosphate
JECFA	The joint FAO/WHO expert committee
MSG	Monosodium glutamate

7.1 Background

Taste and aroma are important factors in determining the flavor of foods. Sweet, salty, sour, bitter, and umami comprise the five basic tastes. Each taste is recognized by specific receptors and is associated with particular neural transmission pathways. However, foods have sensory attributes that cannot be explained by aroma and the five basic tastes alone, such as texture, continuity, complexity, and mouthfulness.

Ueda et al. (1990) investigated the flavoring effects of a diluted extract of garlic that enhanced continuity, mouthfulness, and thickness of taste when added to an umami solution. In that study, they attempted to isolate and identify the key compounds responsible for this effect. First, they prepared an active fraction by cation exchange chromatography using a Duolite C-25 column. The scheme of the preparation is shown in Fig. 7.1. The enhancing effect of continuity, mouthfulness, and

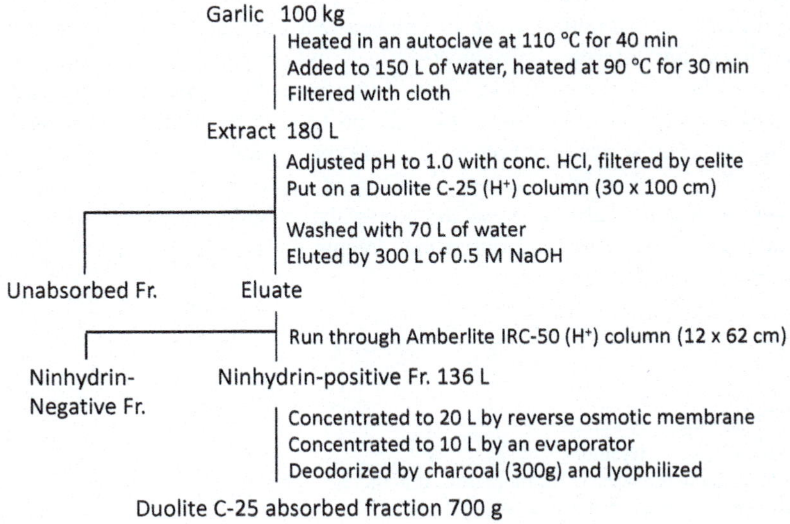

Fig. 7.1 Preparation of Duolite C-25 absorbed fraction from garlic

Table 7.1 Content of each sulfur-containing compound in Duolite C-25 adsorbed fraction

Compound[a]	Content (%)
Alliin	16.0
Cycloalliin	9.2
MeCSO	2.1
GAC	5.3
GACSO	4.7
Glutathione (GSH)	0.2
Cysteine	3.0
Methionine	0.1

[a]*MeCSO* S-methyl-cysteine sulfoxide, *GAC* γ-glutamyl S-allyl-cysteine, *GACSO* γ-glutamyl S-allyl-cysteine sulfoxide

Table 7.2 Effect of each sulfur-containing compound in salty/umami solution

Compound[a]	Enhancement of taste	Other flavor
Alliin	+++	Garlic-like
Cycloalliin	+	
MeCSO	++	Leak-like
GAC	++	Garlic-like
GACSO	+	Garlic-like
Glutathione (GSH)	+++	
Cysteine	+	
Methionine	+	

[a]*MeCSO* S-methyl-cysteine sulfoxide, *GAC* γ-glutamyl S-allyl-cysteine, *GACSO* γ-glutamyl S-allyl-cysteine sulfoxide

thickness of taste to umami solution was contained in the Duolite C-25 absorbed fraction. The components of the Duolite C-25 absorbed fraction were analyzed, and the results are demonstrated in Table 7.1. The results indicated that the major components of the fraction were S-allyl-L-cysteine sulfoxide (alliin), cycloalliin, S-methyl-L-cysteine sulfoxide (MeCSO), γ-glutamyl-S-allyl-cysteine (GAC), γ-glutamyl-S-allyl-cysteine sulfoxide (GACSO), cysteine (Cys), and glutathione (GSH). To compare the sensory effect of the components found in Duolite C-25 absorbed fraction, each component was prepared and added at 0.2% (w/v) to umami solution, and the enhancement of continuity, mouthfulness, and thickness of taste was evaluated by a sensory panel. The results of the sensory evaluation (Table 7.2) revealed that alliin and glutathione exerted the strongest effect among the compounds tested. Because glutathione did not have any other flavor and was relatively stable, Ueda et al. focused on the sensory characteristics of glutathione (Ueda et al. 1997). They investigated the threshold of glutathione in various systems. The results

Table 7.3 Threshold values of glutathione in water and umami solutions

Solution	Threshold value (%)
Water	0.04
MSG 0.05%	0.04
MSG 0.80%	0.02
MSG 3.1%	0.01
MSG 0.05% + IMP 0.05%	0.01

in Table 7.3 show that the threshold in an umami solution (which consists of a solution containing 3.1% MSG and solution containing 0.05% MSG and 0.05% IMP) was much lower than that measured in water. Thus, glutathione has only a minimal flavor in water, but if added to an umami solution, it can substantially enhance the thickness of taste, continuity, and mouthfulness (Ueda et al. 1997). Ueda et al. proposed that substances with these properties should be referred to as *kokumi* substances. Therefore, as cited in Chap. 1, it is considered that *kokumi* substance is considered as one of "*koku*"-imparting factors.

The research group led by Professor Thomas Hofmann at the Technical University of Munich investigated *kokumi* substances from various foodstuffs by sensory evaluation-guided elucidation (sensomics approach). They investigated the sensory effects of various compounds isolated from foods by the panelists who were trained using 5 mM glutathione in model chicken broth for the training of the activity of mouthfulness enhancement and complexity increase (Dunkel et al. 2007). Dunkel et al. isolated γ-glutamyl-leucine (γ-Glu-Leu), γ-glutamyl-valine (γ-Glu-Val), and γ-glutamyl-cysteinyl-β-alanine (γ-Glu-Cys-β-Ala: homoglutathione) as *kokumi* substances from edible beans (*Phaseolus vulgaris* L.). They indicated that the threshold values of these peptides were much lower in salty/umami solution and chicken broth than those in water. For example, threshold of γ-Glu-Cys-β-Ala in salty/umami solution was 0.1 mM while that in water was 3.8 mM (Dunkel et al. 2007). Toelstede et al. (2009) identified γ-glutamyl-glutamic acid (γ-Glu-Glu), γ-glutamyl-glycine (γ-Glu-Gly), γ-glutamyl-glutamine (γ-Glu-Gln), γ-glutamyl-methionine (γ-Glu-Met), γ-glutamyl-leucine (γ-Glu-Leu), γ-glutamyl-valine (γ-Glu-Val), and γ-glutamyl-histidine (γ-Glu-His) as *kokumi* substances from Gouda cheese. They indicated that the threshold values of these peptides were much lower in the reconstituted Gouda cheese extract solution than those in water. For example, the threshold of γ-Glu-His in the reconstituted Gouda cheese extract solution was 0.01 mM, while that in water was 2.5 mM (Toelstede et al. 2009). They also detected γ-glutamyl-alanine (γ-Glu-Ala), γ-glutamyl-phenylalanine (γ-Glu-Phe), γ-glutamyl-tyrosine (γ-Glu-Tyr) in Gouda cheese. Hillmann et al. (2016) identified *kokumi* γ-glutamyl peptides including γ-glutamyl-glutamic acid (γ-Glu-Glu), γ-glutamyl-glycine (γ-Glu-Gly), γ-glutamyl-glutamine (γ-Glu-Gln), γ-glutamyl-methionine (γ-Glu-Met), γ-glutamyl-leucine (γ-Glu-Leu), γ-glutamyl-valine (γ-Glu-Val), and γ-glutamyl-histidine (γ-Glu-His) in Parmesan cheese and detected γ-glutamyl-aspartic acid (γ-Glu-Asp), γ-glutamyl-phenylalanine (γ-Glu-Phe), γ-glutamyl-tyrosine (γ-Glu-Tyr), γ-glutamyl-threonine (γ-Glu-Thr), and γ-glutamyl-lysine (γ-Glu-Lys).

Shibata et al. (2017) attempted to identify the key *kokumi* substances in soybean seeds and identified γ-glutamyl-tyrosine (γ-Glu-Tyr) and γ-glutamyl-phenylalanine (γ-Glu-Phe) as key *kokumi* peptides in soybean seeds. They also reported that the threshold of γ-Glu-Tyr is 0.005 mM in a salty/umami solution, which is much lower than that in water (3.0 mM).

Kokumi substances such as GSH are reported to be perceived via the calcium-sensing receptor (CaSR) in humans (Ohsu et al. 2010). The studies described above indicated that GSH and several other peptides can activate human CaSR, as can several γ-glutamyl peptides, including γ-Glu-Ala, γ-Glu-Val, γ-Glu-Cys, γ-Glu-α-aminobutyryl-Gly (ophthalmic acid), and γ-Glu-Val-Gly, activate human CaSR. Furthermore, these γ-glutamyl peptides possess the characteristics of *kokumi* substances which can modify the five basic tastes (especially sweet, salty, and umami) when added to basic taste solutions or food, even though they have no taste themselves at the concentrations. The CaSR activity of these γ-glutamyl peptides is positively correlated with the sensory activity of *kokumi* substances, suggesting these γ-glutamyl peptides are perceived via CaSR in humans. Among these, γ-Glu-Val-Gly has been reported to be a potent *kokumi* peptide with a sensory activity that is 12.8-fold times greater than that of GSH (Ohsu et al. 2010). Additionally, γ-Glu-Val-Gly is present in several foods such as scallops (Kuroda et al. 2012a), fermented fish sauces (Kuroda et al. 2012b; Miyamura et al. 2016), soy sauces (Kuroda et al. 2013, 2015), fermented shrimp paste (Miyamura et al. 2014), and beer (Miyamura et al. 2015a, b).

These studies indicate that various *kokumi* γ-glutamyl peptides are widely distributed in foods. Table 7.4 summarizes the reported *kokumi* γ-glutamyl peptides and their occurrence in foods.

These previous reports on various *kokumi* peptides suggest that γ-Glu-Val-Gly is a potent *kokumi* peptide. In the next section, we describe the sensory effect of γ-Glu-Val-Gly, compared with those of glutathione (GSH), on commercial chicken soup as assessed using sensory evaluation.

7.2 Sensory Effects of Glutathione and γ-Glutamyl-Valyl-Glycine in Commercial Chicken Soup

7.2.1 Introduction

As mentioned in the previous section, Ohsu et al. (2010) reported that the *kokumi* substances such as GSH are reported to be perceived through the calcium-sensing receptor (CaSR) in humans. To confirm the hypothesis, the CaSR activity and the sensory activity as *kokumi* substances of various γ-glutamyl peptides were investigated (Ohsu et al. 2010). The sensory activity as *kokumi* substance was determined

Table 7.4 *Kokumi* γ-glutamyl peptides found in foods

Peptides	Occurrence in foods[a]
Dipeptides	
γ-Glu-Cys	Gouda cheese[1], Parmesan cheese[2], yeast extract[3]
γ-Glu-Val	Edible beans (*Phaseolus vulgaris*)[4], Gouda cheese[1], Parmesan cheese[2]
γ-Glu-Leu	Edible beans (*Phaseolus vulgaris*)[4], Gouda cheese[1], Parmesan cheese[2]
γ-Glu-Glu	Gouda cheese[1], Parmesan cheese[2]
γ-Glu-Gly	Gouda cheese[1], Parmesan cheese[2]
γ-Glu-Gln	Gouda cheese[1], Parmesan cheese[2]
γ-Glu-Met	Gouda cheese[1], Parmesan cheese[2]
γ-Glu-His	Gouda cheese[1], Parmesan cheese[2]
γ-Glu-Ala	Gouda cheese[1], Parmesan cheese[2]
γ-Glu-Phe	Gouda cheese[1], Parmesan cheese[2], soybean[5]
γ-Glu-Tyr	Gouda cheese[1], Parmesan cheese[2], soybean[5]
Tripeptides	
γ-Glu-Cys-Gly(GSH)	Garlic[6], onion[7], yeast extract[3], scallop[8], wine[8], beef meat[8], chicken meat[8]
γ-Glu-Cys-β-Ala (homoglutathione)	Edible beans (*Phaseolus vulgaris*)[4]
γ-Glu-Val-Gly	Scallop[9], fish sauce[10], soy sauce[11], beer[12], shrimp paste[13]
γ-Glu-Abu-Gly (ophthalmic acid)	Yeast extract
Dipeptide derivatives	
γ-Glu-S-Allyl-Cys	Garlic[6]
γ-Glu-S-Allyl-Cys sulfoxide	Garlic[6]
γ-Glu-S-(1-propenyl)-Cys sulfoxide	Onion[7]

[a]Superscript numbers indicate the following reference. (1) Toelstede et al. (2009), (2) Hillmann et al. (2016), (3) Kuroda et al. (1997), (4) Dunkel et al. (2007), (5) Shibata et al. (2017), (6) Ueda et al. (1990), (7) Ueda et al. (1994), (8) Ueda et al. (1997), (9) Kuroda et al. (2012a), (10) Kuroda et al. (2012b), (11) Kuroda et al. (2013), (12) Miyamura et al. (2015a, b), (13) Miyamura et al. (2014)

by the point of subjective equivalent (PSE) method. The peptides tested were γ-Glu-Ala, γ-Glu-Val, γ-Glu-Cys, γ-Glu-α-aminobutyryl-Gly (ophthalmic acid), γ-Glu-Val-Gly, and glutathione (γ-Glu-Cys-Gly). Of these, γ-Glu-Val-Gly was shown to have the highest sensory activity, which was 12.8-fold times greater than that of glutathione, indicating that γ-Glu-Val-Gly is a potent *kokumi* peptide.

In this part of our present study, the sensory effect of this peptide on commercial chicken soup was compared with that of glutathione (GSH) using sensory evaluation.

7.2.2 *Materials and Methods*

7.2.2.1 Materials

Food additive grade glutathione (GSH, γ-Glu-Cys-Gly) was purchased from Kojin Life Science Co., Ltd. (Tokyo, Japan). The γ-Glu-Val-Gly used in the present study was of food additive grade (FEMA-GRAS No. 4709; JECFA food flavoring No.2123), was obtained from Ajinomoto Co., Inc. (Tokyo, Japan), and was prepared by a chemical synthetic method reported previously (Ohsu et al. 2010). Commercial chicken consommé powder was purchased from Ajinomoto Co., Inc. (Tokyo Japan).

7.2.2.2 Sensory Panel

In this study, 20 trained panelists (all were men; 24–50 years old) participated in the sensory evaluation. All panelists were the employees of Ajinomoto Co., Inc., and were working on the development of foods and are residents in Tokyo metropolitan area, Japan. All of them passed the sensory panel examination conducted using a previously described method (Furukawa 1977).

7.2.2.3 Sensory Evaluation of Commercial Chicken Soup with GSH and γ-Glu-Val-Gly

Chicken soup containing 2% commercial chicken consommé powder (Ajinomoto Co., Inc.) was prepared by dissolving the powder in hot tap water (control soup). GSH (0.02%) and γ-Glu-Val-Gly (0.002%) were then dissolved in the control soup to prepare the sample soups. Approximately 50 mL of each soup in a plastic cup was served with a control soup. The intensities of thickness of taste, continuity, and mouthfulness were evaluated. Thickness of taste was expressed in terms of increased taste intensity at ~5 s after tasting. Continuity was expressed as taste intensity at ~20 s after tasting. Mouthfulness was expressed as reinforcement of the taste sensation throughout the mouth and not just on the tongue. A trained panel of 20 assessors was asked to evaluate the sample solutions on a 5-point rating scale that ranged from −2 (apparently suppressed) to +2 (apparently strong).

7.2.2.4 Statistical Analysis

Statistical analysis was conducted using JMP version 9.0 (SAS Institute, Cary, NC, USA). Data of the sensory evaluation are collected as the means ± standard error. Data were assessed by the paired *t*-test. Differences were considered statistically significant at $p < 0.05$.

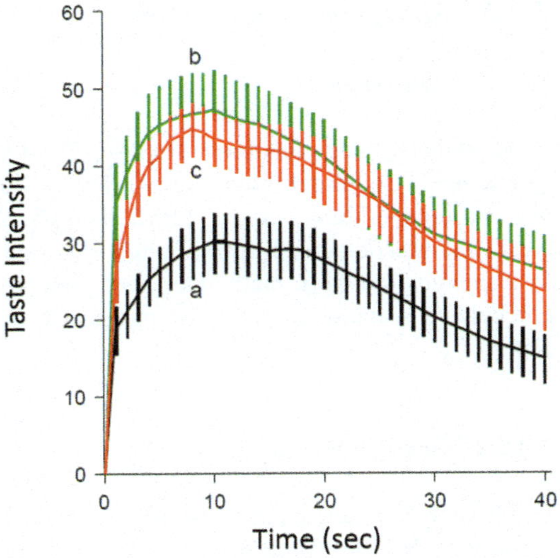

Fig. 7.2 Sensory effect of GSH and γ-Glu-Val-Gly on a commercial chicken soup. The closed bar reveals the effect of 200 ppm (651 μM) GSH. The open bar reveals the effect of 20 ppm (66 μM) γ-Glu-Val-Gly. $*p < 0.05$

7.2.3 Results and Discussion

7.2.3.1 Effects of GSH and γ-Glu-Val-Gly on Commercial Chicken Soup

The examination was performed to evaluate and reveal if there is an enhancement of the continuity, mouthfulness, and thickness of taste when GSH or γ-Glu-Val-Gly was added to a commercial chicken soup. The results in Fig. 7.2 show that both GSH (0.02%) and γ-Glu-Val-Gly (0.002%) enhanced the intensity of continuity, mouthfulness, and thickness of taste at nearly the same level. γ-Glu-Val-Gly enhanced all three characteristics (thickness of taste, continuity, and mouthfulness). γ-Glu-Val-Gly showed significant enhancement at a concentration as low as 0.002%. These results suggest that γ-Glu-Val-Gly is a potent *kokumi* substance. In the following sections, we investigated the effect of γ-Glu-Val-Gly on various foods.

7.3 Effect of *Kokumi* Peptide, γ-Glutamyl-Valyl-Glycine, on the Sensory Characteristics of Chicken Consommé

7.3.1 Introduction

Previously, Ohsu et al. (2010) also reported that adding 0.01% γ-Glu-Val-Gly to 3.3% sucrose solution, 0.9% NaCl solution, and 0.5% monosodium glutamate (MSG) solution significantly enhanced sweetness, saltiness, and umami, respectively. They also reported that adding 0.002% γ-Glu-Val-Gly to chicken consommé prepared from commercial chicken consommé powder significantly enhanced thickness, continuity, and mouthfulness. In that report, sensory evaluation was undertaken with sensory attributes with reference to a method reported previously (Ueda et al. 1990, 1997).

In the present study, we characterized the sensory properties of food with added γ-Glu-Val-Gly, using a descriptive analysis of chicken consommé containing γ-Glu-Val-Gly.

7.3.2 Materials and Methods

7.3.2.1 Preparation of γ-Glu-Val-Gly

The γ-Glu-Val-Gly used in the present study was of food additive grade (FEMA-GRAS No. 4709; JECFA food flavoring No.2123), was obtained from Ajinomoto Co., Inc. (Tokyo, Japan), and was prepared using a chemical synthetic method reported previously (Ohsu et al. 2010).

7.3.2.2 Preparation of Chicken Consommé

The raw materials for chicken consommé are shown in Table 7.5. Minced chicken breast meat, minced chicken leg meat, and egg white were mixed. Then, minced chicken wing meat was added and mixed. The raw materials (except bouillon and water) were mixed in a 60-L aluminum pot. Bouillon (Kisco Co Ltd., Tokyo, Japan) diluted with the same volume of water was added and boiled at between 90 °C and 95 °C for 30 min. After removing the meat, precipitate, and fat, the resulting chicken consommé was freeze-dried (freezing temperature, −24 °C; vacuum, <13 Pa; sample

Table 7.5 Raw materials for the chicken consommé

Materials	Weight (g)
Chicken breast meat (minced)	6818.2
Chicken leg meat (minced)	6818.2
Chicken wing meat (minced)	6818.2
Egg white	1500.0
Fried onion	1687.5
Carrot	562.5
Celery	375.0
Tomato	1406.3
Tomato paste	150.0
Parsley	18.8
Black pepper	5.6
Bouillon (Kisco Co., Inc.)	15000.0
Water	15000.0

temperature, <20 °C) using a freeze dryer (RL-50 MB, Kyowa Vacuum Engineering Co. Ltd., Tokyo, Japan). For sensory evaluation, 5.6 g of the freeze-dried chicken consommé powder and 0.2 g of sodium chloride were dissolved in 100 mL of distilled water and heated to 60 °C and presented to the panelists. Approximately 90 mL of consommé was served in foam cups coded with three-digit random numbers.

7.3.2.3 Selection of the Panel

Eighteen female panelists participated in the sensory evaluation. The age of the panelists was 54.0 ± 8.8 (mean ± standard deviation) years old. All the panelists resided in the San Francisco Bay Area, CA, USA. Screening of the panelists was conducted in three phases: phone screening of applicants, on-site acuity testing, and face-to-face interviews with advanced acuity testing.

7.3.2.4 Training of the Panel

All of the panelists were broadly trained in sensory descriptive analysis to evaluate aromas, flavors, textures, and appearance across a wide range of consumer products. This training was conducted for approximately 3 days per week for 3 months, during which the panelists expanded their food sensory vocabularies, learned to use a 15-point scale to rate attribute intensities and evaluated a wide variety of foods. For example, the sweetness intensity scale was anchored with several concentrations of sucrose in water and the intensity of "sweet aromatic" was anchored with several concentrations of vanilla in milk. The panelists in training refined their skills by participating in practice tests using many different types of products. After each test, they were given detailed feedback while retesting the products to help them improve their performance. After this training was complete, the panelists were registered as members of the Descriptive Panel of the National Food Laboratory and began to participate in the descriptive analysis of various kinds of foods.

7.3.2.5 Selection of the Sensory Attributes

During this session, panelists evaluated samples of chicken consommé with and without γ-Glu-Val-Gly to understand the effect of γ-Glu-Val-Gly. A panel leader led the group in discussion on the differences and similarities between the samples. They developed a list of sensory attributes that described the products' sensory characteristics, focusing on attributes believed to be influenced by γ-Glu-Val-Gly. Each sample was tested at least twice during this orientation session. The panelists also developed new attributes such as "total chicken/meaty flavor," "bone and marrow flavor," "roasted flavor," "richness," "tongue-coating," and "salivating." Overall, the panelists defined the 17 sensory attributes listed in Table 7.6. The panelists practiced rating the samples on the list so that they were prepared to begin data collection.

Table 7.6 Definition and reference samples for the descriptive attributes of chicken consommé

Sensory attributes	Definitions	Reference samples and intensity
Total flavor	The total intensity of all of the flavors of the sample including basic tastes	Kitchen Basics Chicken Broth (6)
Total chicken/ meaty flavor	The flavor intensity reminiscent of cooked chicken meat	Kitchen Basics Chicken Broth (5)
Chicken flavor	The flavor intensity reminiscent of cooked chicken	Kitchen Basics Chicken Broth (5)
Bone and marrow flavor	The character associated with chicken bones, particularly the marrow of chicken bones	NR[a]
Roasted flavor	The total flavor intensity that is reminiscent of roasted chicken and/or vegetables	Swanson's Chicken Broth (6)
Total vegetable flavor	The total flavor intensity of vegetables such as carrots, green vegetables, and herbs in the broth	Kitchen Basics Chicken Broth (5)
Richness	The degree to which the flavor characters of the sample are harmonized, balanced, and blend well together as opposed to being spiky or striking out	NR
Salty	One of the basic taste, common to sodium chloride	0.2% sodium chloride in water (2)
		0.5% sodium chloride in water (5)
Umami	One of the basic taste, common to MSG. The taste and mouth-filling sensation of compounds such as glutamates, that is, savory, brothy, meaty, rich, full, and complex, common to many foods such as soy sauce, stocks, ripened cheese (especially parmesan), shellfish (crab, lobster, scallops, clams), mushrooms (especially porcini), ripe tomatoes, cashews, and asparagus	Kitchen Basics Chicken Broth (2)
		0.5% MSG in Kitchen Basics Chicken Broth (3.5)
Viscosity	The degree to which the samples in viscous in mouth from thin to thick	Water (1), heavy whipping cream (6)
Mouthfulness	The perception that the sample fills the whole mouth, is blooming, or is growing, a full-bodied sensation when the sample is held in the mouth	Kitchen Basics Chicken Broth (1.5)
		0.5% MSG in Kitchen Basics Chicken Broth (3)

(continued)

Table 7.6 (continued)

Sensory attributes	Definitions	Reference samples and intensity
Mouth-coating	The degree to which there is a leftover residue, a slick, powdery, or fatty coating or film in the mouth that is difficult to clear	0.5% MSG in water (4)
		Half and half (5)
Tongue-coating	The *degree* to which there is a leftover residue, a slick, powdery, or fatty coating or film on the tongue that is difficult to clear	0.5% MSG in water (3)
Total trigeminal	The intensity of the total sensation, including numbing, burning, tingling, or irritation, impaired on the soft tissues of the oral cavity, particularly the tongue	Wintergreen Breath Saver (NS[b])
		0.5% MSG in water (5)
Salivating	The degree to which the sample caused a perceived increase in salivation	NR
Swelling perception of soft tissue	The feeling of swelling of the soft tissue in the oral cavity, specifically the cheeks and lips, reminiscent of the perception of swelling produced by antithetic treatments at a dental office, but without a distinct numbing effect	0.5% MSG in water (4)
Total aftertaste	The total aftertaste intensity after 5 s of all flavor notes within the sample	NR

[a]*NR* no reference
[b]*NS* not scored

7.3.2.6 Procedure for Sensory Evaluation

For the evaluation of chicken consommé, panelists held the product in the mouth for 10 s, expectorated, and then rated flavor, texture/mouthfeel, and aftertaste attributes. They then completed the rating for each attribute (samples with and without 5 ppm γ-Glu-Val-Gly) on a 15-point line scale. The sample serving order was balanced, with each sample being presented approximately an equal number of times in each position for each test. Human sensory analyses were conducted following the spirit of the Helsinki Declaration, and informed consent was obtained from all panelists. The experimental protocol was approved by the ethic board of Institute of Food Sciences & Technologies, Ajinomoto.

7.3.2.7 Analyses of Free Amino Acids and 5′-Nucleotides in Chicken Consommé

Free amino acids were determined using a model L-8800 amino acid analyzer (Hitachi Corp., Tokyo, Japan) with lithium citrate buffer (PF-series for nonhydrolyzed amino acid analysis; Mitsubishi Chemical, Tokyo, Japan). The contents of 5′-nucleotides were determined by HPLC equipped with a Hitachi #3013 column with detection at 254 nm.

7.3.2.8 Statistical Analyses

Statistical analyses were conducted using JMP version 9.0 (SAS Institute Inc., Cary, NC, USA). The data were collected as the means ± standard deviation. Data were assessed by the paired t-test. The data was considered to be significant when the confidence level was more than 95%.

7.3.3 Results and Discussion

7.3.3.1 Sensory Attributes for Chicken Consommé

During the project-specific orientation session, the panelists developed 17 attributes shown in Table 7.6. Regarding the attributes related to chicken flavor, since many words related to the chicken flavor were proposed during the project-specific orientation session, three attributes, "total chicken/meaty flavor," "bone and marrow flavor," and "roasted flavor," were added to the list. Total chicken/meaty flavor was defined as the flavor intensity reminiscent of cooked chicken meat, bone and marrow flavor was defined as the character associated with chicken bones, particularly the marrow of chicken bones, and roasted flavor was defined as the total flavor intensity that is reminiscent of roasted chicken and/or vegetables. Additionally, because the coating sensation was well recognized when the panelists evaluated the chicken consommé with γ-Glu-Val-Gly during the project-specific panel orientation session, the attributes "mouth-coating" and "tongue-coating" were added to the list. "Mouth-coating" was defined as the degree to which there is a leftover residue, a slick, powdery, or fatty coating or film on the mouth that is difficult to clear. "Tongue-coating" was defined as the degree to which there is a leftover residue, a slick, powdery, or fatty coating or film on the tongue that is difficult to clear. Overall, the panelists defined the 17 sensory attributes for chicken consommé listed in Table 7.2: nine taste and flavor attributes (total flavor, total chicken/meaty flavor, chicken flavor, bone and marrow flavor, roasted flavor, total vegetable flavor, richness, salty, and umami), seven texture/mouthfeel attributes (viscosity, mouthfulness, mouth-coating, tongue-coating, salivating, total trigeminal, and swelling perception of soft tissue), and one aftertaste (total aftertaste). Definitions of these sensory attributes and the references are shown in Table 7.6.

7.3.3.2 Sensory Characteristics of Chicken Consommé with Added γ-Glu-Val-Gly

The sensory characteristics of chicken consommé with or without γ-Glu-Val-Gly are shown in Table 7.7 and Fig. 7.3. The addition of γ-Glu-Val-Gly at 5 ppm significantly enhanced the intensity of umami and mouthfulness at a 99% confidence level. Furthermore, the addition of this peptide significantly enhanced the intensity

Table 7.7 Sensory characteristics of chicken consommé with added γ-Glu-Val-Gly

Sensory attributes	Control consommé	Consommé with γ-Glu-Val-Gly	Changed value	95% confidence interval	99% confidence interval	Significance
Total flavor	6.13 ± 0.72	6.31 ± 0.68	0.18 ± 0.60	0.28	0.36	N.S.
Total chicken/meaty flavor	5.26 ± 0.61	5.41 ± 0.59	0.14 ± 0.59	0.27	0.36	N.S.
Chicken flavor	4.82 ± 0.55	4.88 ± 0.78	0.06 ± 0.69	0.32	0.42	N.S.
Bone and marrow flavor	2.42 ± 0.85	2.63 ± 0.98	0.21 ± 1.14	0.53	0.69	N.S.
Roasted flavor	3.19 ± 1.03	3.12 ± 1.03	−0.07 ± 0.82	0.38	0.50	N.S.
Total vegetable flavor	3.56 ± 0.75	3.78 ± 0.81	0.22 ± 0.64	0.30	0.39	N.S.
Richness	4.01 ± 0.81	4.27 ± 0.94	0.27 ± 0.79	0.36	0.48	N.S.
Salty	2.73 ± 0.48	2.87 ± 0.73	0.13 ± 0.57	0.26	0.35	N.S.
Umami	2.84 ± 0.65	3.28 ± 0.67	0.43 ± 0.66	0.30	0.40	**
Viscosity	2.06 ± 0.65	2.22 ± 0.59	0.16 ± 0.40	0.18	0.24	#
Mouthfulness	2.47 ± 0.70	2.92 ± 0.73	0.45 ± 0.69	0.32	0.42	**
Mouth-coating	2.67 ± 0.66	2.94 ± 0.65	0.27 ± 0.56	0.26	0.34	*
Tongue-coating	2.56 ± 0.82	2.72 ± 0.82	0.17 ± 0.68	0.32	0.42	N.S.
Salivating	2.42 ± 0.85	2.58 ± 0.83	0.16 ± 1.06	0.49	0.64	N.S.
Total trigeminal	2.76 ± 0.87	2.98 ± 0.73	0.23 ± 0.84	0.39	0.51	N.S.
Swelling perception of soft tissue	2.78 ± 0.79	2.87 ± 0.68	0.09 ± 0.76	1.48	0.46	N.S.
Total aftertaste	4.46 ± 0.60	4.54 ± 0.66	0.08 ± 0.64	0.29	0.38	N.S.

Data was shown as means ± standard deviations
*significant at a 95% confidence level; **, significant at a 99% confidence level; #, trend at a 90% confidence level
N.S. not significant

of mouth-coating at a 95% confidence level. Adding this peptide at 5 ppm did not significantly change the intensity of the other attributes. A recent study has suggested that *kokumi* peptides such as GSH and γ-Glu-Val-Gly enhance the intensity of umami if they are added to 0.5% MSG solution (Ohsu et al. 2010), an observation consistent with the present study. Additionally, in the descriptive analysis, umami has been defined as not only the "taste of MSG" but also as "the mouth-filling sensation of compounds such as glutamates that is savory, brothy, meaty, rich, full, and complex, which is common to many foods such as soy sauce, stocks, ripened cheese, shellfish, mushrooms, ripened tomatoes, cashews, and asparagus" (Table 7.6).

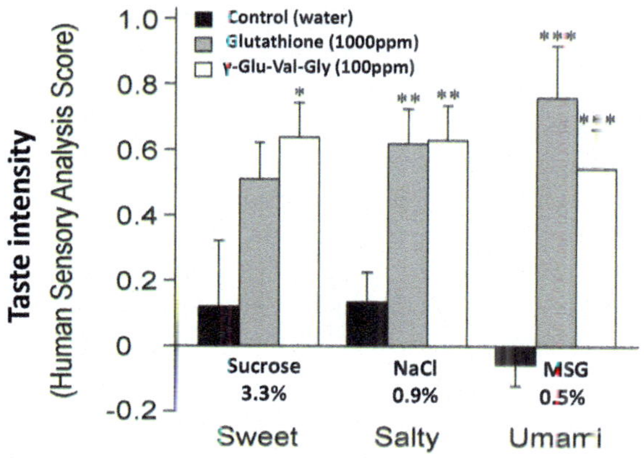

Fig. 7.3 Graphical representation of the sensory characteristics of chicken consommé with added γ-Glu-Val-Gly. A blue fine line indicates the mean scores of the control consommé. A red line indicates the mean scores of the consommé with added 5 ppm of γ-Glu-Val-Gly
∗, significant at a 95% confidence level; ∗∗, significant at a 99% confidence level; #, trend at a 90% confidence level

Therefore, it appears that the enhancement of umami in chicken consommé includes the enhancement of sensations such as richness and complexity. The present results also suggest that γ-Glu-Val-Gly also enhanced mouthfulness. A previous study demonstrated that adding γ-Glu-Val-Gly at 20 ppm to chicken soup significantly enhanced mouthfulness which is consistent with the present study (Ohsu et al. 2010). Regarding other γ-glutamyl peptides, it has been reported that several *kokumi* γ-glutamyl peptides enhanced mouthfulness in food systems. Ueda et al. (1997) reported that the addition of GSH (γ-Glu-Cys-Gly) enhanced the intensity of mouthfulness in model beef meat extract. In addition, Ohsu et al. (2010) also reported that the addition of GSH enhanced the intensity of mouthfulness in chicken soup. Furthermore, it has been reported that γ-glutamyl peptides such as γ-Glu-Val, γ-Glu-Leu, and γ-Glu-Cys-β-Ala found as *kokumi*-active peptides in edible beans enhanced mouthfulness when they were added to chicken broth (Dunkel et al. 2007). In addition, it has been reported that γ-Glu-Glu, γ-Glu-Gly, γ-Glu-His, γ-Glu-Gln, γ-Glu-Met, and γ-Glu-Leu were the key components which impart long-lasting mouthfulness of matured Gouda cheese (Toelstede et al. 2009). From these observations, it is demonstrated that many *kokumi* γ-glutamyl peptides enhance the intensity of mouthfulness.

Interestingly, the present study has revealed that the addition of γ-Glu-Val-Gly at 5 ppm significantly enhanced the intensity of mouth-coating. It has been generally known that mouth-coating sensation is evoked by the addition of hydrocolloids such as xanthan gum and locust bean gum, carrageenan (Arocas et al. 2010; Flett et al. 2010), and fat-containing food materials such as dairy fat emulsion (Flett et al. 2010). However, several studies have reported that low-molecular-weight compounds enhanced the intensity of mouth-coating. Dawid and Hofmann reported that 1,2-dithiolan-4-carboxylic acid 6-D-glucopyranoside ester exhibited a buttery mouth-coating sensation (Dawid and Hofmann 2012). Additionally, the same research group demonstrated that polyphenolic compounds such as vanillin, vanillin-related compounds, americanin A, and 4′,6′-dihydroxy-3′,5′-dimethoxy-[1,1′-biphenyl]-3-carboxaldehyde from cured vanilla beans exhibited a velvety mouth-coating sensation (Schwarz and Hofmann 2009). Furthermore, it has been reported that the flavon-3-ol glycosides such as kaempferol glycosides, quercetin glycosides, myricetin glycosides, and apigenin glycoside from black tea induced a mouth-coating sensation (Scharbert et al. 2004). Despite these observations, there have been no reports of a peptide which exhibited the mouth-coating sensation. Therefore, this is the first report which has demonstrated the mouth-coating effect of peptides. Although the viscosity of consommé did not change significantly by adding 5 ppm of γ-Glu-Val-Gly (data not shown), an enhancement of the mouth-coating sensation was observed. The mechanism of this enhancement will be clarified in our future studies.

In the present study, the addition of γ-glutamyl-valyl-glycine enhanced the intensity of umami, mouthfulness, and mouth-coating. Conversely, monosodium glutamate (MSG), the representative umami compound, also enhances the intensity of mouthfulness and sensation related to mouth-coating (Yamaguchi and Kimizuka 1979). Previously, it has been demonstrated that γ-glutamyl-valyl-glycine enhanced the umami intensity when it was added to 0.3% MSG solution (Ohsu et al. 2010). In addition, as shown in Table 7.8, the analysis of the chemical components revealed that chicken consommé contained glutamic acid (51.1 mg/dl) and IMP (21.3 mg/dl), and these concentrations of umami components were sufficient to evoke the umami sensation (Yamaguchi and Kimizuka 1979). Therefore, one possibility is that the enhancement of mouthfulness and mouth-coating by γ-glutamyl-valyl-glycine was caused by the enhancement of umami. The unique characteristic of *kokumi* compounds is that unlike umami compounds, *kokumi* compounds have no taste themselves and can only enhance taste sensation when added to foods. For that reason, *kokumi* compounds likely enhance sensations such as mouthfulness and continuity in sweet foods. In our recent study, it was observed that γ-glutamyl-valyl-glycine enhanced aftertaste, oiliness, and mouthfulness in reduced-fat peanut butter (Miyamura et al. 2015a, b). This result suggests that *kokumi* compounds can be used both in savory and sweet foods, while umami compounds can be used mainly in savory foods because of the characteristic umami taste. Further detailed studies are necessary to clarify the mechanism of the enhancement of the mouthfulness and mouth-coating sensation by γ-glutamyl-valyl-glycine.

The addition of γ-Glu-Val-Gly significantly enhanced the intensity of umami, mouthfulness, and mouth-coating in chicken consommé. The results suggest that adding γ-Glu-Val-Gly can improve the flavor and mouthfeel of chicken consommé. Consumer preferences for chicken consommé are currently being investigated to determine whether this compound can be used to enhance the desirable character of food.

Table 7.8 Contents of free amino acids and 5′-nucleotide in chicken consommé

Component	Content (mg/dl)	Component	Content (mg/dl)
Amino acids			
Taurine	74.9	Isoleucine	7.0
Aspartic acid	17.5	Leucine	12.4
Threonine	27.5	Tyrosine	9.9
Serine	20.3	Phenylalanine	8.0
Glutamic acid	51.1	Lysine	17.0
Glycine	15.8	Histidine	6.6
Alanine	23.6	Arginine	23.7
Valine	10.2	Hydroxyproline	2.5
Methionine	4.7	Proline	10.0
5′-Nucleotide			
5′-IMP	21.3		
5′-GMP	n.d.		

∗*n.d.* not detected

7.3.4 Conclusions and Implications

In the present study, the sensory characteristics of chicken consommé with 5.0 ppm added γ-Glu-Val-Gly were investigated using descriptive analysis. Chicken consommé containing γ-Glu-Val-Gly had significantly stronger "umami" and "mouthfulness" (mouth-filling sensation) characteristics than the control sample at a 99% confidence level and significantly stronger "mouth-coating" characteristic than a control at a 95% confidence level. These data indicated that a *kokumi* peptide, γ-Glu-Val-Gly, can enhance umami, mouthfulness, and mouth-coating in chicken consommé. From these results, it was suggested that the addition of γ-Glu-Val-Gly can improve the flavor and mouthfeel of chicken consommé.

7.4 Effect of *Kokumi* Peptide, γ-Glutamyl-Valyl-Glycine, on the Sensory Characteristics of Reduced-Fat Peanut Butter

7.4.1 Introduction

In the previous section, we described the effect of γ-Glu-Val-Gly on the sensory characteristics of chicken consommé. Adding γ-Glu-Val-Gly to chicken consommé significantly enhanced thickness of taste (taste enhancement assessed ~5 sec after tasting), continuity (taste intensity at 20 sec after tasting), and mouthfulness (the reinforcement of taste sensation throughout the mouth and not just on the tongue) (Ohsu et al. 2010). It is generally known that these sensations are evoked by the addition of fat-containing food materials such as dairy fat emulsion (Flett et al. 2010).

The increased health issues associated with obesity have resulted in the development and commercialization of various low-fat foods and reduced-fat foods. However, in general, the palatability of low-fat foods and reduced-fat foods is lower than that of full-fat foods (McClements and Demetriades 1998). Previous studies using descriptive sensory analysis have shown that low-fat Lyon-style sausages (17% fat and 10% fat sausages; full-fat sausage contains 25% fat) have decreased meat flavor, aftertaste meat flavor, juiciness, and greasiness, while they have increased spiciness, aftertaste spiciness, firmness, and coarseness (Tomaschunas et al. 2013). Low-fat yogurt (0.1% fat; full-fat yogurt contains 3.5% fat) shows a decreased score with respect to creaminess, which is defined as sensations of mouth-coating, and shows increased scores of firmness and homogeneity (Pimentel et al. 2013). Low-fat ice cream (4.0% fat; full-fat ice cream contains 10.0% fat) shows lower scores in thickness, smoothness, creaminess, mouth-coating, cooked sugar flavor, milky flavor, condensed milk flavor, and milky aftertaste (Liou and Grun 2007), while low-fat cheddar cheese (14.5% fat: full-fat cheddar cheese contains 34.7% fat) scores lower with respect to milk fat flavor and brothy flavor (Amelia et al. 2013). Low-fat foods mainly lack texture; therefore, thickeners such as gums, starch, and modified starch can be used to resolve this issue. However, low-fat foods with such additives still show lower palatability than that of full-fat foods.

In the present study, we aimed to clarify whether addition of γ-Glu-Val-Gly changed the sensory characteristics of reduced-fat foods and low-fat foods. We investigated the effect of γ-Glu-Val-Gly on reduced-fat peanut butter.

7.4.2 *Materials and Methods*

7.4.2.1 Preparation of γ-Glu-Val-Gly

The γ-Glu-Val-Gly used in the present study was of food additive grade (FEMA-GRAS No. 4709; FEMA: Flavor and Extract Manufacturers' Association; JECFA food flavoring No. 2123; JECFA: Joint FAO/WHAO Expert Committee on Food Additives), was obtained from Ajinomoto Co., Inc. (Tokyo, Japan), and was prepared by chemical synthesis as reported previously (Ohsu et al. 2010).

7.4.2.2 Preparation of Reduced-Fat and Full-Fat Peanut Butter

The raw materials for reduced-fat peanut butter (30% fat content) and full-fat peanut butter model (50% fat content) are shown in Table 7.9. Regarding the preparation of the reduced-fat peanut butter, the emulsifiers were mixed with peanut paste in an aluminum pot at 30 °C with later addition of cream by stirring. Sugar and salt solubilized in water were added, stirred, and heated at 40 °C for 5 minutes. Regarding

Table 7.9 Raw materials for the low-fat peanut butter and full-fat model peanut butter

Materials	Low-fat (wt%)	Full-fat model (wt%)
Peanut paste	55.0	55.0
Salad oil	0.0	21.0
Salt	1.0	1.0
Sugar (granulated)	6.2	6.2
Cream	5.0	5.0
Emulsifier (sugar ester; HLB:15)	2.0	0.5
Emulsifier (glyceryl monostearate; HLB:4)	0.0	2.0
Xanthan gum	0.0	0.5
Water	30.8	8.8

the preparation of the full-fat model peanut butter, the emulsifiers were mixed with peanut paste and salad oil in an aluminum pot at 30 °C with later addition of cream by stirring. Sugar and salt solubilized in water were added, stirred, and heated at 40 °C for 5 min. For preparation of reduced-fat peanut butter with γ-Glu-Val-Gly, γ-Glu-Val-Gly was added by dissolving in water with sugar and salt and then prepared according to the methods mentioned above. Prepared peanut butter samples were packed in glass bottles and stored at 4 °C until sensory evaluation.

7.4.2.3 Selection of the Sensory Panel

In this study, 29 panelists (17 men and 12 women; 28.8 ± 5.0 years old, mean ± standard deviation) participated in the sensory evaluation. All panelists were the employees of Ajinomoto Shanghai Food Research and Technology Center and were working on the development of foods. They were Chinese and were residents of Shanghai city. In addition, all of them passed the sensory panel examination conducted using a previously described method (Furukawa 1977). For the comparison between the reduced-fat peanut butter and full-fat model, 20 panelists (9 men and 11 women; 27.6 ± 3.6 years old, mean ± standard deviation) participated in the sensory evaluation. For the investigation of the effect of γ-Glu-Val-Gly, 19 panelists (13 men and 6 women; 29.9 ± 5.3 years old, mean ± standard deviation) participated in the evaluation.

7.4.2.4 Selection of the Sensory Attributes

The panelists evaluated samples of the reduced-fat peanut butter and full-fat peanut butter. A panel leader led the group in discussion on the differences and similarities between the samples. To describe the sensory characteristics of the products, the panelists developed a list of nine sensory attributes: peanut flavor, saltiness, sweetness, bitterness, thickness of taste, aftertaste, continuity of taste, smoothness, and oiliness. The panelists practiced rating the samples on the list so that they were prepared to begin data collection.

7.4.2.5 Procedure for Sensory Evaluation

The sensory evaluation was conducted between 10:00 am and 11:30 am in partitioned booths of sensory evaluation room, with temperature set at 25 °C via air-conditioning. For evaluation of the peanut butter samples, 10 g of sample was spread on one piece of bread (10 g), which was cut into four pieces. The panelists held each piece of bread with peanut butter in the mouth, evaluated the taste, and rated each attribute. They rinsed their mouths with commercial mineral water between the samples. They completed the rating for each attribute on a 3-point linear scale: −1.0, apparently weaker than the control; 0, same as the control; and 1.0, apparently stronger than the control. For comparison between the reduced-fat sample and full-fat model, half the panelists evaluated the full-fat model using a reduced-fat sample as the control, and the other half evaluated the reduced-fat sample using a full-fat model as the control. Combination of the samples was randomized and balanced. Human sensory analyses were conducted following the spirit of the Helsinki Declaration, and informed consent was obtained from all panelists. The experimental procedure was approved by the Ethics Board of the Institute of Food Sciences & Technologies, Ajinomoto Co., Inc. (Tokyo, Japan).

7.4.2.6 Statistical Analysis

Statistical analysis was conducted using JMP version 9.0 (SAS Institute, Cary, NC, USA). The data were collected as the means ± standard error. Data were assessed by the paired *t*-test. Differences were considered statistically significant at $p < 0.05$.

7.4.3 Results and Discussion

7.4.3.1 Sensory Attributes

During the group discussion, panelists listed up the words, selected the attributes, and came to a consensus with respect to which sensation the attribute expressed. The panelists developed 10 attributes: peanut flavor, saltiness, sweetness, bitterness, thickness of taste (enhancement of taste intensity with maintaining the balance of taste), aftertaste (total aftertaste intensity after 5 sec of all flavor notes within the sample), continuity of taste (total taste intensity at 20 sec after tasting), smoothness, and oiliness.

7.4.3.2 Comparison of Reduced-Fat and Full-Fat Peanut Butter

The comparison between reduced-fat peanut butter and full-fat model peanut butter is shown in Table 7.10. The full-fat model peanut butter had higher scores for peanut flavor, thickness of taste, aftertaste, continuity of taste, and oiliness than low-fat peanut butter. No significant difference in saltiness, sweetness, bitterness, smoothness, and viscous sensation was observed between the low-fat sample and full-fat model. We consider that the fat enhanced the above sensory character in peanut butter. In other words, we considered that peanut flavor, thickness of taste, aftertaste, continuity of taste, and oiliness were the sensory functions of fat in peanut butter (Table 7.11). Peanut flavor was defined as flavor intensity reminiscent of peanut. Thickness of taste was defined as enhancement of taste intensity with maintaining the balance of taste. Aftertaste was defined as total taste intensity at 5 sec after tasting. Continuity was defined as total taste intensity at 20 sec after tasting. Oiliness was defined as the oily mouth-coating sensation.

Table 7.10 Result of the comparison test between low-fat peanut butter and full-fat model peanut butter

Sensory attributes	Score of full-fat model	Significance
Peanut flavor	0.24 ± 0.05	**
Saltiness	0.04 ± 0.04	N.S.
Sweetness	−0.03 ± 0.05	N.S.
Bitterness	−0.10 ± 0.05	N.S.
Thickness of taste	0.15 ± 0.06	*
Aftertaste	0.16 ± 0.04	**
Continuity of taste	0.14 ± 0.04	**
Smoothness	0.10 ± 0.08	N.S.
Viscosity	−0.01 ± 0.09	N.S.
Oiliness	0.23 ± 0.07	**

Data was shown as means ± standard errors
$*p < 0.05$, $**p < 0.01$
N.S. not significant

Table 7.11 Sensory attributes and their definition used in evaluation on the effect of γ-Glu-Val-Gly on low-fat peanut butter

Sensory attributes	Definition
Peanut flavor	The flavor intensity reminiscent of peanut
Thickness of taste	The enhancement of taste intensity with maintaining the balance of taste
Aftertaste	The total teste intensity at 5 sec after tasting
Continuity of taste	The total teste intensity at 20 sec after tasting
Oiliness	The oily mouth-coating sensation

7.4.3.3 Effect of Addition of γ-Glu-Val-Gly on Reduced-Fat Peanut Butter

We next evaluated the effect of γ-Glu-Val-Gly on the sensory characteristics of reduced-fat peanut butter. For this, reduced-fat peanut butter with the addition of 40 ppm (132 μM) γ-Glu-Val-Gly was evaluated for peanut flavor, thickness of taste, aftertaste, continuity of taste, and oiliness. The results of the sensory evaluation are shown in Table 7.12 and Fig. 7.4. The addition of γ-Glu-Val-Gly significantly enhanced the intensity of thickness of taste, aftertaste, and oiliness. These results indicate that the addition of γ-Glu-Val-Gly increased some sensations that were lacking in the reduced-fat peanut butter. These results also indicate that addition of γ-Glu-Val-Gly can be used to improve the flavor of reduced-fat peanut butter.

Previous studies have shown that several reduced-fat foods and low-fat foods lacked the sensations related to "thickness of taste," "aftertaste," and "oiliness." For example, low-fat Lyon-style sausages (10% fat sausages; full-fat sausage contains

Table 7.12 Effect of γ-Glu-Val-Gly on low-fat peanut butter

Sensory attributes	Score of sample with γ-Glu-Val-Gly	Significance
Peanut flavor	0.06 ± 0.05	NS
Thickness of taste	0.13 ± 0.04	**
Aftertaste	0.14 ± 0.05	*
Continuity of taste	0.09 ± 0.05	NS
Oiliness	0.09 ± 0.04	*

Data was shown as means ± standard errors
$*p < 0.05$, $**p < 0.01$
NS not significant

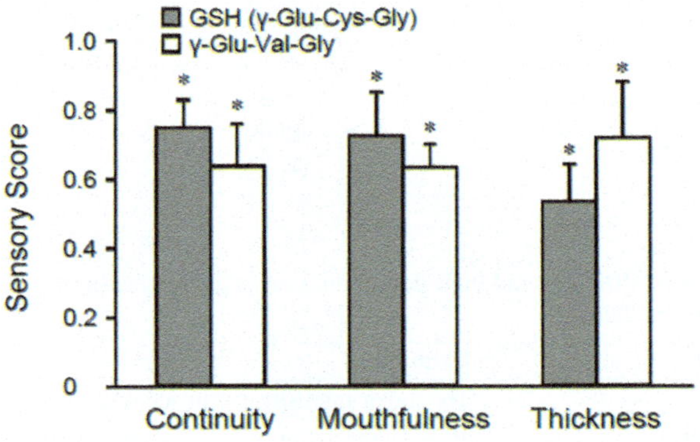

Fig. 7.4 Graphical representation of the sensory characteristics of reduced-fat peanut butter with added γ-Glu-Val-Gly. A blue line indicates the mean scores of the control reduced-fat peanut butter. A red line indicates the mean scores of the reduced-fat peanut butter with added 40 ppm of γ-Glu-Val-Gly
$*p < 0.05$, $**p < 0.01$

25% fat) have decreased meat flavor, aftertaste meat flavor, juiciness, and greasiness, while they have increased spiciness, aftertaste spiciness, firmness, and coarseness (Tomaschunas et al. 2013). In that study, low-fat Lyon-type sausage containing inulin and citrus fiber was prepared to improve the flavor and texture, and the result indicated that aftertaste meat flavor, firmness, juiciness, and coarseness of the low-fat sausage containing inulin and citrus fiber is not significantly different from the full-fat sausage. However, meat flavor and greasiness were still lower than those of full-fat sausage, and spiciness and aftertaste spiciness were still higher than those of full-fat sausage (Tomaschunas et al. 2013). Low-fat yogurt (0.1% fat; full-fat yogurt contains 3.5% fat) shows a decreased score with respect to creaminess, which is defined as sensations of mouth-coating, and shows increased scores of firmness and homogeneity (Pimentel et al. 2013). In that study, low-fat yogurt containing long-chain inulin was prepared to improve the flavor and texture, and the result indicated that firmness of the low-fat yogurt with long-chain inulin was not significantly different from the full-fat yogurt. However, creaminess was still lower than that of full-fat yogurt, and homogeneity was still higher than that of full-fat yogurt (Pimentel et al. 2013). These previous studies suggested that ingredients such as inulin and citrus fiber could improve some characteristics which are lacking in low-fat foods; however, they also suggested that these ingredients could not improve the characteristics such as meat flavor, greasiness, and creaminess. From the results shown in this section, it is possible that γ-Glu-Val-Gly can be used to improve the flavor of other reduced-fat foods containing the ingredients inulin and dietary fibers as fat replacer. To evaluate this possibility, it is necessary to conduct a preference test using a consumer panel, and this test is now in progress in our laboratory. The effect of γ-Glu-Val-Gly on other reduced-fat foods will be shown in the following sections.

7.4.4 Conclusions and Implications

In this study, the effect of a *kokumi* peptide, γ-Glu-Val-Gly, on the flavor of reduced-fat peanut butter was investigated. The results indicated that the addition of γ-Glu-Val-Gly significantly enhanced the intensities of thickness of taste, aftertaste, and oiliness. These results show that addition of γ-Glu-Val-Gly increased the sensory attributes that were lacking in reduced-fat peanut butter, suggesting that the addition of this peptide can improve the flavor of reduced-fat peanut butter.

7.5 Effect of *Kokumi* Peptide, γ-Glutamyl-Valyl-Glycine, on the Sensory Characteristics of Reduced-Fat Custard Cream

7.5.1 Introduction

In the previous section, the effect of γ-Glu-Val-Gly on the sensory characteristics of reduced-fat peanut butter was investigated. The results indicated that the addition of γ-Glu-Val-Gly (40 ppm; 162 μM) significantly enhanced the intensities of thickness of taste, aftertaste, and oiliness. It was interesting that the addition of γ-Glu-Val-Gly enhanced the oiliness which is defined as the oily mouth-coating sensation. These sensations are evoked by the addition of fat-containing food materials such as dairy fat emulsion (Flett et al. 2010).

As mentioned previously, increased obesity in the population has spurred the development and commercialization of various reduced-fat foods; however, the palatability of reduced-fat foods is generally lower than that of full-fat foods (McClements and Demetriades 1998). Among the foods containing dairy cream, several studies have been performed on the sensory characteristics of low-fat ice cream, and the previous reports indicated that low-fat ice cream show lower score in flavor and texture-related attributes such as creaminess and smoothness (Patel et al. 2010; Abdel-Haleem and Awad 2015; Azari-Anpar et al. 2017; de Souza Fernandes et al. 2017; Sharma et al. 2017; Zhang et al. 2018; Guo et al. 2018). Descriptive analysis on the low-fat ice cream has shown that low-fat ice cream (4.0% fat; full-fat ice cream contains 10.0% fat) shows lower scores in thickness, smoothness, creaminess, mouth-coating, cooked sugar flavor, milky flavor, condensed milk flavor, and milky aftertaste (Liou and Grun 2007). In the previous section, it was indicated that the addition of γ-Glu-Val-Gly, a potent *kokumi* peptide, to reduced-fat peanut butter increased the intensity of thickness of taste, aftertaste, and oiliness. This result suggests that the addition of γ-Glu-Val-Gly can improve the flavor of reduced-fat foods.

In the present study, we aimed to clarify whether addition of γ-Glu-Val-Gly can change the flavor of various reduced-fat foods. Hence, we investigated the effect of γ-Glu-Val-Gly on the sensory characteristics of reduced-fat custard cream.

7.5.2 Materials and Methods

7.5.2.1 Preparation of γ-Glu-Val-Gly

The γ-Glu-Val-Gly used in the present study was of food additive grade (FEMA-GRAS No. 4709; FEMA: Flavor and Extract Manufacturers' Association; JECFA food flavoring No. 2123; JECFA: Joint FAO/WHAO Expert Committee on Food Additives), was obtained from Ajinomoto Co., Inc. (Tokyo, Japan), and was prepared by chemical synthesis as reported previously (Ohsu et al. 2010).

7.5.2.2 Preparation of Reduced-Fat and Full-Fat Custard Cream

The raw materials for reduced-fat custard cream (3.6% fat content) and full-fat custard cream model (12.0% fat content) are shown in Table 7.13. Each custard cream was prepared as follows. The raw materials were mixed by a food processer Bamix M-300 (ESGE AG, Switzerland) for 1 min, heated at 95 °C for 30 min, screened through a mesh, and cooled using iced water to room temperature. The prepared custard cream samples were stored in a refrigerator at 4 °C until sensory evaluation.

7.5.2.3 Selection of the Sensory Panel

In this study, 29 panelists (17 men and 12 women; 28.8 ± 5.0 years old, mean ± standard deviation) participated in the sensory evaluation. All panelists were employees of Ajinomoto Shanghai Food Research and Technology Center and were working in food development. All panelists were Chinese and were residents of Shanghai city. In addition, all of the panelists passed the sensory panel examination conducted using a previously described method (Furukawa 1977). For the comparison of reduced-fat and full-fat custard cream, 20 panelists (9 men and 11 women; 27.6 ± 3.6 years old, mean ± standard deviation) participated in the sensory evaluation. For the investigation of the effect of γ-Glu-Val-Gly, 19 panelists (13 men and 6 women; 29.9 ± 5.3 years old, mean ± standard deviation) participated in the evaluation.

Table 7.13 Raw materials for reduced-fat custard cream and full-fat model custard cream

Materials	Reduced-fat (wt%)	Full-fat model (wt%)
Fresh cream	0.0	19.0
Milk	0.0	45.5
Skim milk	6.0	0.0
Egg yolk (20% sugar added)	14.5	14.5
Sugar	8.0	10.0
Flour	3.0	3.0
Vanilla beans	0.1	0.1
Starch	0.5	0.0
Inulin cream	24.3	0.0
Water	43.6	7.9

7.5.2.4 Selection of the Sensory Attributes

Panelists evaluated samples of the reduced-fat custard cream and full-fat custard cream model. A panel leader led the group in discussion on the differences and similarities between the samples. They developed a list of sensory attributes that described the sensory characteristics of the products. The panelists developed 10 attributes: thickness of taste, continuity of taste, sweetness, aftertaste, egg flavor, vanilla flavor, mildness, and oiliness. The panelists practiced rating the samples on the list so that they were prepared to begin data collection.

7.5.2.5 Procedure for Sensory Evaluation

The sensory evaluation was conducted between 10:00 am and 11:30 am in the partitioned booth at 25 °C in an air-conditioned sensory evaluation room. For evaluation of the custard cream samples, 10 g of sample was put on a plastic spoon, and the panelists held the sample in the mouth, evaluated the taste, and rated each attribute. They rinsed their mouths with commercial mineral water between the samples. They completed the rating for each attribute on a 3-point linear scale: −1.0, apparently weaker than the control; 0, same as the control; and 1.0, apparently stronger than the control. For comparison between the reduced-fat sample and full-fat model, half the panelists evaluated the full-fat model using a reduced-fat sample as the control, and the other half evaluated the reduced-fat sample using a full-fat model as the control. The serving order of samples was randomized and balanced. Human sensory analyses were conducted following the spirit of the Helsinki Declaration, and informed consent was obtained from all panelists. The experimental procedure was approved by the Ethics Board of the Institute of Food Sciences & Technologies, Ajinomoto.

7.5.2.6 Statistical Analysis

Statistical analysis was conducted using JMP version 9.0 (SAS Institute, Cary, NC, USA). All data of the sensory evaluation were presented as the means ± standard error. Data were analyzed using paired *t*-test. Differences were considered statistically significant at $p < 0.05$.

7.5.3 Results and Discussion

7.5.3.1 Sensory Attributes

During the group discussion, panelists listed up the words, selected the attributes, and came to a consensus with respect to which sensation the attribute expressed. Finally, the panelists developed nine attributes: thickness of taste (the enhancement of taste intensity with maintaining the balance of taste), continuity of taste (taste intensity assessed at 20 sec after tasting), sweetness, aftertaste (total taste intensity 5 sec after tasting of all flavor notes within the sample), egg flavor, vanilla flavor, and oiliness.

7.5.3.2 Comparison of Reduced-Fat and Full-Fat Custard Cream

Comparison between reduced-fat custard cream and full-fat custard cream model is shown in Table 7.14. The full-fat custard cream scored significantly higher scores for thickness of taste, continuity, aftertaste, vanilla flavor, and cream flavor than did reduced-fat custard cream. No significant difference in saltiness, sweetness, bitterness, smoothness, and viscous sensation was observed between the reduced-fat samples and full-fat samples. We consider that the fat enhanced the sensory characteristics such as thickness of taste, continuity, aftertaste, vanilla flavor, and cream flavor in custard cream. In other words, we considered that thickness of taste, continuity, aftertaste, vanilla flavor, and cream flavor were the main sensory functions of fat in custard cream. The sensory attributes and their definition are shown in Table 7.15. Thickness of taste was defined as the enhancement of taste intensity

Table 7.14 Result of the comparison test between low-fat custard cream and full-fat model custard cream

Sensory attributes	Score of full-fat model	Significance
Thickness of taste	0.09 ± 0.05	#
Continuity of taste	0.10 ± 0.05	#
Sweetness	−0.03 ± 0.05	N.S.
Aftertaste	0.12 ± 0.06	*
Egg flavor	0.09 ± 0.06	N.S.
Vanilla flavor	0.16 ± 0.05	**
Cream flavor	0.18 ± 0.06	**
Viscosity	−0.01 ± 0.09	N.S.
Oiliness	0.09 ± 0.06	N.S.

Data was shown as means ± standard errors
#$p < 0.1$, *$p < 0.05$, **$p < 0.01$
N.S. not significant

Table 7.15 Sensory attributes and their definition used in evaluation on the effect of γ-Glu-Val-Gly on low-fat custard cream

Sensory attributes	
Thickness of taste	The enhancement of taste intensity with maintaining the balance of taste
Continuity of taste	The total teste intensity at 20 sec after tasting
Aftertaste	The total teste intensity at 5 sec after tasting
Vanilla flavor	The flavor intensity reminiscent of vanilla
Cream flavor	The flavor intensity reminiscent of cream

with maintaining the balance of taste. Continuity was defined as the total taste intensity at 20 s after tasting, and aftertaste was defined as the total taste intensity at 5 s after tasting. Vanilla flavor was defined as the flavor intensity reminiscent of vanilla. Cream flavor was defined as the flavor intensity reminiscent of cream.

7.5.3.3 Effect of γ-Glu-Val-Gly on Reduced-Fat Custard Cream

To clarify the effect of γ-Glu-Val-Gly on the sensory character of reduced-fat custard cream, reduced-fat custard cream with 40 ppm γ-Glu-Val-Gly was evaluated. The results of the sensory evaluation are shown in Table 7.16 and Fig. 7.5. Addition of γ-Glu-Val-Gly significantly enhanced the intensities of thick flavor ($p < 0.05$) and tended to enhance the intensity of continuity of taste and aftertaste ($p < 0.1$). These results demonstrated that the addition of γ-Glu-Val-Gly increased some sensations that were lacking in the reduced-fat custard cream, suggesting that addition of this peptide can be used to improve the flavor of reduced-fat custard cream.

Among the foods containing dairy cream, several studies have been performed on the sensory characteristics of low-fat ice cream, and the previous reports indicated that low-fat ice cream show lower score in flavor and texture-related attributes such as creaminess and smoothness (Abdel-Haleem and Awad 2015; Azari-Anpar et al. 2017; de Souza Fernandes et al. 2017; Sharma et al. 2017; Zhang et al. 2018). To improve the sensory characteristics of low-fat ice cream, various kinds of fat replacers, such as sago powder (Patel et al. 2010), hulless barley flour and barley β-glucan (Abdel-Haleem and Awad 2015), resistant starch and maltodextrin (Azari-Anpar et al. 2017), cassava derivatives (de Souza Fernandes et al. 2017), citrus fiber (Zhang et al. 2018), and citrus pectin (Zhang et al. 2018), and nanobacterial cellulose/soy protein isolate complex gel (Guo et al. 2018) have been tested. Patel et al. (2010) reported that the sensory characteristics such as flavor, body, and texture and melting quality of the low-fat mango ice cream (2.4% fat; full-fat mango ice cream contains 10.0% fat) containing sago powder were not significantly different from those of the full-fat mango ice cream. Reported that the use of resistant starch and maltodextrin can improve the sensory characteristics such as color, texture, and flavor. de Souza Fernandes et al. (2017) reported that the use of cassava bagasse and

Table 7.16 Effect of γ-Glu-Val-Gly on low-fat custard cream

Sensory attributes	Score of sample with γ-Glu-Val-Gly	Significance
Thickness of taste	0.086 ± 0.033	*
Continuity of taste	0.070 ± 0.032	*
Aftertaste	0.063 ± 0.036	#
Vanilla flavor	0.026 ± 0.043	NS
Cream flavor	0.071 ± 0.042	NS

Data was shown as means ± standard errors
*$p < 0.05$, #$p < 0.1$
NS not significant

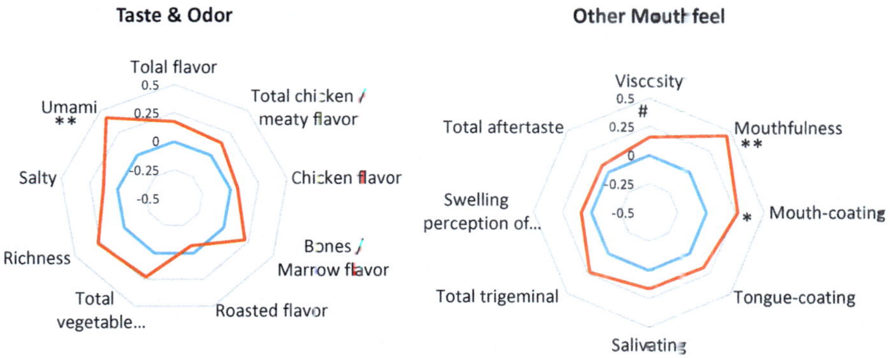

Fig. 7.5 Graphical representation of the sensory characteristics of reduced-fat custard cream with added γ-Glu-Val-Gly. A blue line indicates the mean scores of the control reduced-fat custard cream. A red line indicates the mean scores of the reduced-fat custard cream with added 40 ppm of γ-Glu-Val-Gly
#$p < 0.1$, *$p < 0.05$

maltodextrin can improve the sensory characteristics such as flavor, color, appearance, texture, and taste. Zhang et al. reported that the scores of appearance, flavor, taste, and smoothness of reduced-fat ice cream (45% fat decrease) containing citrus pectin were not significantly different from those of the full-fat ice cream, but the score of mouth-coating of reduced-fat ice cream containing citrus pectin was significantly lower than that of the full-fat ice cream (Zhang et al. 2018). Guo et al. (2018) reported that the texture profiles such as hardness, springiness, chewiness, cohesiveness, and resilience of the low-fat ice cream containing nanobacterial cellulose/soy protein isolate complex gel were not significantly different from those of the full-fat ice cream. However, in these previous studies, the effect of fat replacer on the detailed sensory characteristics has not been revealed.

Descriptive sensory analysis on the low-fat ice cream has shown that low-fat ice cream (4.0% fat; full-fat ice cream contains 10.0% fat) shows lower scores in thickness, smoothness, creaminess, mouth-coating, cooked sugar flavor, milky flavor, condensed milk flavor, and milky aftertaste (Liou and Grun 2007). In that study, low-fat ice cream containing fat mimetics such as microparticulated whey protein concentrate and polydextrose powder was prepared to improve the flavor and texture, and the result indicated that the scores of thickness, smoothness, creaminess, cooked sugar flavor, and condensed milk flavor were not significantly different from those of full-fat ice cream. However, milky flavor and milky aftertaste of low-fat ice cream containing fat mimetics were still lower than those of full-fat ice cream (Liou and Grun 2007).

From the results shown in this section, it is possible that γ-Glu-Val-Gly can be used to improve the flavor of other reduced-fat foods containing fat mimetics or fat replacers. To evaluate this possibility, it is necessary to conduct a preference test using a consumer panel, and this test is now in progress in our laboratory. The effect of γ-Glu-Val-Gly on other reduced-fat foods will be shown in the following sections.

7.5.4 *Conclusions and Implications*

In this study, the effect of a *kokumi* peptide, γ-Glu-Val-Gly, on the flavor of reduced-fat custard cream was investigated. The results indicated that the addition of γ-Glu-Val-Gly significantly enhanced the intensities of thickness of taste and tended to enhance the continuity of taste and aftertaste. These results demonstrated that addition of γ-Glu-Val-Gly increased some sensations that were lacking in the reduced-fat custard cream, suggesting that addition of the peptide could improve the flavor of reduced-fat custard cream.

7.6 Effect of *Kokumi* Peptide, γ-Glutamyl-Valyl-Glycine, on the Sensory Characteristics of Reduced-Fat French Dressing

7.6.1 *Introduction*

In the previous sections, the effect of γ-Glu-Val-Gly on the sensory characteristics of reduced-fat peanut butter and reduced-fat custard cream was investigated. The results indicated that the addition of γ-Glu-Val-Gly (40 ppm; 162 μM) to reduced-fat peanut butter significantly enhanced the intensities of thickness of taste, aftertaste, and oiliness. In addition, γ-Glu-Val-Gly (40 ppm; 162 μM) enhanced the thickness and continuity of taste when added to reduced-fat custard cream.

As mentioned in the previous section, the problem of the increase in the obese population has resulted in various kinds of reduced-fat foods being developed and commercialized. However, in general, the palatability of reduced-fat foods is lower than that of full-fat foods (McClements and Demetriades 1998). In the previous section, it was indicated that the addition of γ-Glu-Val-Gly, a potent *kokumi* peptide, to reduced-fat peanut butter increased the intensity of thickness of taste, aftertaste, and oiliness. This result suggests that the addition of γ-Glu-Val-Gly can improve the flavor of reduced-fat foods.

Salad dressing including French dressing, mayonnaise, and Thousand Island dressing is a common food product worldwide. The oil content of salad dressing is high; it is generally more than 40% in French dressing and Thousand Island dressing and is more than 65% in traditional mayonnaise. Hence, it is generally regarded as a high-fat and high-caloric food. Many studies have been done on the preparation of low-fat salad dressing, especially in mayonnaise-type dressing. Preparation and evaluation of low-fat mayonnaise, containing polysaccharide gums (Wendin et al.

1997), whey protein isolate and pectin (Liu et al. 2007; Sun et al. 2018), mucilage from plants (Aghdaei et al. 2014; Bernardino-Nicanor et al. 2015; Fernandes and Mellado 2018), or an extruded flour paste (Roman et al. 2015), and of low-fat mayonnaise produced via double emulsions (Yildrim et al. 2016) have been reported. Preparation and evaluation of French dressing containing sucrose polyester emulsifier (Mellies et al. 1985) is also reported.

In the present study, we aimed to clarify whether addition of γ-Glu-Val-Gly changed the flavor by palatability of various reduced-fat foods. We investigated the effect of γ-Glu-Val-Gly on reduced-fat French dressing.

7.6.2 Materials and Methods

7.6.2.1 Preparation of γ-Glu-Val-Gly

The γ-Glu-Val-Gly used in the present study was of food additive grade (FEMA-GRAS No. 4709; FEMA: Flavor and Extract Manufacturers' Association; JECFA food flavoring No. 2123; JECFA: Joint FAO/WHAO Expert Committee on Food Additives) obtained from Ajinomoto Co., Inc. (Tokyo, Japan), and was prepared by chemical synthesis as reported previously (Ohsu et al. 2010).

7.6.2.2 Preparation of Reduced-Fat and Full-Fat French Dressing

The raw materials for reduced-fat French dressing (15.0% fat content) and full-fat French dressing (37.5% fat content) are shown in Table 7.17. Each French dressing sample cream was prepared as follows. The raw materials except for soybean oil were mixed by a handy food processer Bamix M-300 (ESGE AG, Switzerland) for 2 min, heated at 85 °C for 10 min, and cooled by iced water to room temperature. Then, soybean oil was added and homogenized at 10,000 rpm for 2 min using a laboratory-scale homogenizer (Labolution, Primix Corp., Hyogo, Japan). Prepared French dressing samples were stored in a refrigerator at 4 °C until sensory evaluation.

7.6.2.3 Sensory Panel

In this study, 29 panelists (17 men and 12 women; 28.8 ± 5.0 years old, mean ± standard deviation) participated in the sensory evaluation. All panelists were the employees of Ajinomoto Shanghai Food Research and Technology Center and were working on the development of foods. They were Chinese and were residents of Shanghai city. In addition, all of them passed the sensory panel examination conducted using a previously described method (Furukawa 1977). Nineteen panelists

Table 7.17 Raw materials for the reduced-fat French dressing and full-fat French dressing model

Materials	Reduced-fat (wt%)	Full-fat model (wt%)
High-fructose corn syrup	28.20	28.20
Vinegar	23.10	23.10
Soybean oil	15.00	37.50
Salt	2.20	2.20
Hydroxypropyl distarch phosphate	1.60	1.60
Paprika powder	1.10	1.10
Citric acid	0.40	0.40
Xanthan gum	0.60	0.28
Emulsifier (polyoxyethylene sorbitan monostearate)	0.40	0.40
Roast garlic powder	0.40	0.40
Tomato flavor	0.60	0.60
Lemon flavor	0.05	0.05
Food coloring agent	2.40	2.40
Water	23.95	1.77

(13 men and 6 women; 29.9 ± 5.3 years old, mean ± standard deviation) participated in the evaluation.

7.6.2.4 Selection of the Sensory Attributes

Panelists evaluated samples of the reduced-fat French dressing and full-fat French dressing model. A panel leader led the group in discussion on the differences and similarities between the samples. They developed a list of sensory attributes that described the sensory characteristics of the products.

7.6.2.5 Procedure for Sensory Evaluation

The sensory evaluation was conducted between 10:00 am and 11:30 am in the partitioned booth at 25 °C in an air-conditioned sensory evaluation room. For evaluation of the samples, 15 g of French dressing sample was mixed with 25 g of cut lettuce, and the panelists held the sample in the mouth, evaluated the taste, and rated each attribute. They rinsed their mouths with commercial mineral water between the samples. They completed the rating for each attribute on a 3-point linear scale: −1.0, apparently weaker than the control; 0, same as the control; and 1.0, apparently stronger than the control. Combination of the samples was randomized and balanced. Human sensory analyses were conducted following the spirit of the Helsinki Declaration, and informed consent was obtained from all panelists. The experimental procedure was approved by the Ethics Board of the Institute of Food Sciences & Technologies, Ajinomoto.

7.6.2.6 Statistical Analysis

Statistical analysis was conducted using JMP version 9.0 (SAS Institute, Cary, NC, USA). The data were collected as the means ± standard error. Data were assessed by the paired *t*-test. The data were considered to be significant at $p < 0.05$.

7.6.3 Results and Discussion

7.6.3.1 Selection of Sensory Attributes

Comparison between the full-fat French dressing and reduced-fat French dressing was performed by all panelists, and the sensory attributes which lacked in the reduced-fat French dressing were selected. During the group discussion, panelists listed up the words, selected the attributes, and made a consensus of the sensation which the attribute expressed. Finally, the panelists developed five attributes: spice flavor, thickness of taste, (the enhancement of taste intensity with maintaining the balance of taste), initial taste, aftertaste (the total aftertaste intensity after 5 s of all flavor notes within the sample), and smoothness. The spice flavor was defined as the flavor intensity reminiscent to spice, especially paprika and garlic The thickness of taste was defined as the enhancement of taste intensity with maintaining the balance of taste. The initial taste was defined as the total taste intensity after 1 s of all flavor notes within the sample. The aftertaste was defined as the total taste intensity after 5 s of all flavor notes within the sample. The smoothness was defined as the oily smooth sensation. The sensory attributes and their definition are shown in Table 7.18.

7.6.3.2 Effect of Addition of γ-Glu-Val-Gly on Reduced-Fat French Dressing

To clarify the effect of γ-Glu-Val-Gly on the sensory character of reduced-fat French dressing, reduced-fat French dressing with 40 ppm γ-Glu-Val-Gly was evaluated. The results of the sensory evaluation are shown in Table 7.19 and Fig. 7.6. Addition of γ-Glu-Val-Gly significantly enhanced the intensities of aftertaste ($p < 0.05$) and tended to enhance the intensity of thickness of taste and initial taste ($p < 0.1$). These results demonstrated that the addition of γ-Glu-Val-Gly increased some sensations that were lacking in the reduced-fat French dressing, suggesting that addition of the peptide can be used for flavor improvement in reduced-fat French dressing.

As mentioned previously, preparation and evaluation of low-fat mayonnaise, containing polysaccharide gums (Wendin et al. 1997), whey protein isolate and pectin (Liu et al. 2007; Sun et al. 2018), mucilage from plants (Aghdaei et al. 2014; Bernardino-Nicanor et al. 2015; Fernandes and Mellado 2018), or an extruded flour paste (Roman et al. 2015), and of low-fat mayonnaise produced via double emulsions (Yildrim et al. 2016) have been reported. Although many of these studies have

Table 7.18 Sensory attributes and their definition used in evaluation on the effect of γ-Glu-Val-Gly on the reduced-fat French dressing

Sensory attributes	Definition
Spice flavor	The flavor intensity reminiscent of spice, especially paprika and garlic
Thickness of taste	The enhancement of taste intensity with maintaining the balance of taste
Initial taste	The total teste intensity at 1 sec after tasting
Aftertaste	The total teste intensity at 5 sec after tasting
Smoothness	The oily smooth sensation

Table 7.19 Effect of γ-Glu-Val-Gly on reduced-fat French dressing

Sensory attributes	Score of sample with γ-Glu-Val-Gly	Significance
Spice flavor	0.05 ± 0.04	NS
Thickness of taste	0.07 ± 0.04	#
Initial taste	0.07 ± 0.04	#
Aftertaste	0.08 ± 0.03	*
Smoothness	0.01 ± 0.04	NS

$*p < 0.05$; $\#p < 0.1$
Data was shown as means ± standard errors
NS not significant

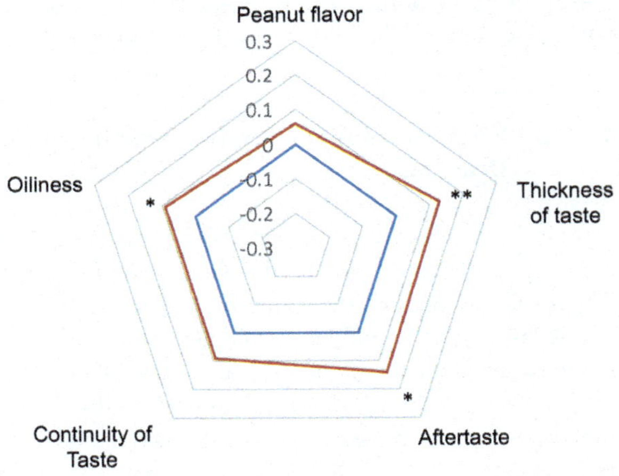

Fig. 7.6 Graphical representation of the sensory characteristics of reduced-fat French dressing with added γ-Glu-Val-Gly. A blue line indicates the mean scores of the control reduced-fat French dressing. A red line indicates the mean scores of the reduced-fat French dressing with added 40 ppm of γ-Glu-Val-Gly **#p < 0.1, *p < 0.05**

been performed focusing on the physicochemical property, however, sensory evaluation of the low-fat mayonnaise containing fat mimetics or fat replacer has been done in several studies. Liu et al. (2007) investigated the rheological, texture, and sensory properties of low-fat mayonnaise with various fat mimetics and reported that the low-fat mayonnaise (40% fat; full-fat mayonnaise contains 80% fat) containing pectin weak gel shows almost similar scores of odor texture, taste, and acceptability as those of full-fat mayonnaise. Aghdaei et al. (2014) reported that the sensory scores such as odor, texture, taste, and mouthfeel of the low-fat mayonnaise (46.8% fat; full-fat mayonnaise contains 78% fat) containing Lafarzeh seed mucilage were not significantly different from those of the full-fat mayonnaise. Su et al. (2010) reported that the sensory scores such as aroma, taste, and greasiness of low-fat mayonnaise (36.5% fat; full-fat mayonnaise contains 73% fat) containing xanthan gum and guar gum was not significantly different from those of the full-fat mayonnaise. However, in these previous studies, the effect of the fat replacer or fat mimetics on the detailed sensory characteristics has not been revealed.

From the results shown in this section, it is possible that γ-Glu-Val-Gly can be used to improve the flavor of salad dressing containing fat replacers or fat mimetics.

7.6.4 Conclusions and Implications

In this study, the effect of a *kokumi* peptide, γ-Glu-Val-Gly, on the flavor of reduced-fat French dressing was investigated. The results indicated that the addition of γ-Glu-Val-Gly significantly enhanced the intensities of thickness of taste and tended to enhance the continuity of taste and aftertaste. These results demonstrated that addition of γ-Glu-Val-Gly increased some sensations that were lacking in the reduced-fat French dressing, suggesting that addition of the peptide could improve the flavor of reduced-fat French dressing.

7.7 Distribution of a *Kokumi* Peptide, γ-Glu-Val-Gly, in Various Fermented Foods and the Possibility of Its Contribution to the Sensory Quality of Fermented Foods

7.7.1 Abstract

A potent *kokumi* peptide, γ-glutamyl-valyl-glycine (γ-Glu-Val-Gly), was identified and quantified in various fermented foods using high-performance liquid chromatography-tandem mass spectrometry (LC/MS/MS). This peptide was present in fermented seasoning such as fish sauces, soy sauces, and fermented shrimp pastes. Among the brewed alcoholic beverages, γ-Glu-Val-Gly was detected in beer. In addition, the content of γ-Glu-Val-Gly positively correlated with the quality grade of dark-colored soy sauce (correlation coefficient $\rho = 0.810$, $p < 0.05$). These results suggest that γ-Glu-Val-Gly is widely distributed in fermented foods and that the content of this peptide can be an indicator of sensory quality of fermented foods.

7.7.2 Introduction

Recent studies have indicated that *kokumi* substances such as glutathione are perceived through the calcium-sensing receptor (CaSR) in humans (Ohsu et al. 2010; Maruyama et al. 2012). These studies have confirmed that glutathione can activate human CaSR and that several γ-glutamyl peptides can also activate the CaSR, and these peptides possess the characteristics of *kokumi* substances, which modify the five basic tastes, especially sweet, salty, and umami, when they are added to basic taste solutions or food, even though these substances have no taste themselves at the concentrations tested (Ueda et al. 1997; Dunkel et al. 2007; Ohsu et al. 2010). Among these *kokumi* peptides, γ-Glu-Val-Gly has been reported to be a potent *kokumi* peptide (Ohsu et al. 2010). In our studies, the distribution of γ-Glu-Val-Gly in various foods was investigated. Because the contents of γ-Glu-Val-Gly in foods were very low, a new method for the determination and quantification of this peptide using LC/MS/MS followed by derivatization with 6-aminoquinoyl-N-hydroxysuccinimidyl-carbamate (AQC) reagent was developed (Kuroda et al. 2012a, b, 2013) by modifying the previously reported method (Armenta et al. 2010). In the present study, the presence and quantities of γ-Glu-Val-Gly in various fermented foods are examined. In addition, possibility of the contribution of γ-Glu-Val-Gly to the sensory quality of fermented foods is also discussed.

7.7.3 Materials and Methods

7.7.3.1 Materials

γ-Glu-Val-Gly used in the present study was prepared by chemical synthesis as reported previously (Ohsu et al. 2010). Uniformly ^{15}N-labeled L-arginine (Arg-UN) and L-proline-d7 (Pro-d7) were obtained from Isotec (Tokyo, Japan). An AccQ Fluor reagent kit was purchased from Waters (Milford, MA, USA). Other reagents are of analytical grade. Commercial fish sauces, soy sauces, fermented shrimp pastes, and beer were purchased from the local market.

7.7.3.2 Sample Preparation and Derivatization Procedure

Samples were filtered through a 0.45-μm syringe filter (25-mm GD/X disposable filter, Whatman PLC, Maidstone, UK). The filtrates were then passed through an Amicon Ultra Centrifugal Filter (regenerated cellulose 10,000 MWCO, Millipore, USA) at 7500 g and 4 °C for 15 min. Derivatization with AQC was conducted using the AccQ Fluor reagent kit (Millipore) according to the manufacturer's protocol.

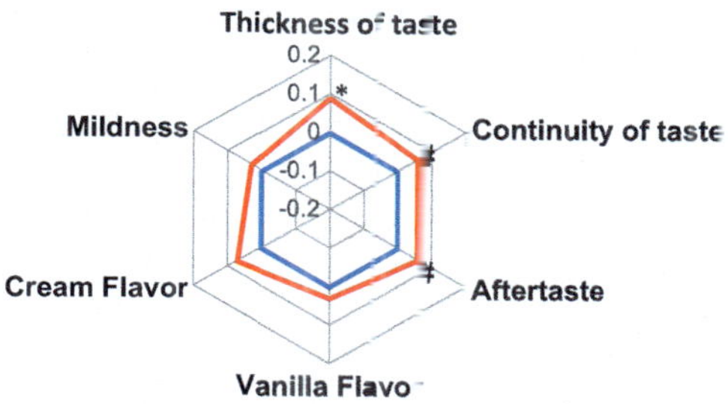

Fig. 7.7 Typical mass spectrum obtained from the AQC-derivative of standard γ-Glu-Val-Gly. The assignment of each fragment ion is also indicated

7.7.3.3 Identification and Quantification of AQC-Derivatized γ-Glu-Val-Gly

Identification and quantification was conducted as reported previously (Kuroda et al. 2012b, 2013; Miyamura et al. 2014, 2015a, b). Separation of AQC-derivatized γ-Glu-Val-Gly was conducted by HPLC using octadecyl silica column. The peak of AQC-derivatized γ-Glu-Val-Gly was identified by the multiple reaction monitoring (MRM) method with six MRM transition channels. The combination of precursor/product ions (Q1/Q3) and the collision energies (CE (V)) were 474.2/171.2 (51 V), 474.2/145.3 (30 V), 474.2/300.3 (30 V), 474.2/229.4 (20 V), 474.2/304.0 (20 V), and 474.2/72.1 (50 V). AQC-derivatized γ-Glu-Val-Gly in the samples was quantified by the internal standard method using the most sensitive channel (474.2/171.2; Q1/Q3). Internal standards (Arg-UN and Pro-d7) were monitored in the MRM transition channels at 349.0/171.1 (Q1/Q3) and 293.0/171.1 (Q1/Q3), respectively. The typical mass spectrum of AQC-derivatized γ-Glu-Val-Gly is indicated in Fig. 7.7.

7.7.3.4 Statistical Analysis

Correlation between the γ-Glu-Val-Gly content and quality grade of soy sauce was analyzed by a Spearman's rank correlation test performed using STAT-VIEW version 5.0 software.

7.7.4 Results and Discussion

7.7.4.1 Distribution of γ-Glu-Val-Gly in Various Fermented Foods

The content of this peptide in 17 brands of commercial fish sauce from Southeast Asia, East Asia, and Europe was quantified, and the results were shown in Table 7.20. The contents in Vietnamese fish sauce (*nuoc mum*) ranged from 10.4 mg/L to 12.6 mg/L and the average was 11.6 mg/L, while those in Thai fish sauce (*nampra*) ranged from 1.2 to 3.1 mg/L, and the average content was 2.3 mg/L. Among the two

Table 7.20 Characteristics of fish sauces and the contents of γ-Glu-Val-Gly in fish sauces

Samples	Country of origin	Raw materials	Contents of γ-Glu-Val-Gly (mg/L)
Nampra A	Thailand	Anchovy, salt	2.7
Nampra B	Thailand	Anchovy, salt, sugar	1.2
Nampra C	Thailand	Anchovy, salt	3.1
Nampra D	Thailand	Anchovy, salt, sugar	2.0
Nampra E	Thailand	Anchovy, salt, sugar	2.3
Nuoc mum A	Vietnam	Anchovy, salt	12.0
Nuoc mum B	Vietnam	Anchovy, salt	12.6
Nuoc mum C	Vietnam	Anchovy, salt	12.3
Nuoc mum D	Vietnam	Anchovy, salt	10.6
Nuoc mum E	Vietnam	Anchovy, salt	10.4
Patis	Philippines	Mackerel, salt	<LOQ
Yu-lu A	China	Fish, salt	<LOQ
Yu-lu B	China	Fish, salt	1.1
Myoruchi extract	Korea	Sardine, salt	<LOQ
Kanari extract	Korea	Sand lance, salt	<LOQ
Shottsuru A	Japan	Sandfish, salt	2.8
Shottsuru B	Japan	Sandfish, salt	<LOQ
Yoshiru	Japan	Sardine, salt	<LOQ
Ikanago-shoyu	Japan	Sand lance, salt	0.5
Garum	Italy	Anchovy, salt	0.4

LOQ limit of quantification

Chinese fish sauce (*yu-lu*), one contained γ-Glu-Val-Gly at 1.1 mg/L. In addition, two of the four Japanese fish sauce contained γ-Glu-Val-Gly at 2.8 and 0.5 mg/L (average content: 0.8 mg/L, n = 4). Furthermore, Italian fish sauce (*garum*) contained γ-Glu-Val-Gly at 0.4 mg/L. These results suggest that γ-Glu-Val-Gly is contained in various fish sauces. γ-Glu-Val-Gly was distributed in commercial Japanese soy sauces at concentrations ranging from 1.5 to 6.1 mg/L as shown in Table 7.21 (Kuroda et al. 2012b). The contents of dark-colored soy sauce ranged from 3.1 to 6.1 mg/L (n = 6), while those in light-colored soy sauce ranged from 3.4 to 3.7 mg/L (n = 2). The content of the white soy sauce sample was 1.5 mg/L (n = 1). These results suggest that γ-Glu-Val-Gly is distributed in soy sauces. As shown in Table 7.22, commercial fermented shrimp paste condiments from Southeast Asia (n = 3) contained this peptide at concentrations ranging from 0.9 to 5.2 mg/kg (Miyamura et al. 2014). Next, the contents of γ-Glu-Val-Gly in various brews including wine, rice wine (*sake*), and beer were analyzed. The analyses indicated that γ-Glu-Val-Gly was present in all of the beer samples (n = 8) at concentrations ranging from 0.08 to 0.18 mg/L as shown in Table 7.23 (Miyamura et al. 2015a, b). In top-fermented beer (n = 2), the content of this peptide was 0.09 mg/L and 0.11 mg/L (average content = 0.11 mg/L), and the bottom-fermented beer (n = 6) contained this peptide in the range between 0.08 and 0.15 mg/L (average content = 0.12 mg/L, n = 6). However, the peptide was not detected in any of the wine or rice wine samples (Miyamura et al. 2015a, b). The analytical data was

Table 7.21 The contents of γ-Glu-Val-Gly in commercial soy sauces

Samples	Grade	Raw materials	Contents of γ-Glu-Val-Gly (mg/L)
Dark-colored soy sauce A	Super	Soybean, wheat, salt	5.2
Dark-colored soy sauce B	Super	Soybean, wheat, salt	5.3
Dark-colored soy sauce C	Ordinary	Defatted soybean, wheat, salt, soybean, ethanol	4.3
Dark-colored soy sauce D	Ordinary	Defatted soybean, wheat, salt, soybean, ethanol	3.1
Dark-colored soy sauce E	Ultrasuper	Defatted soybean, wheat, salt, soybean	6.1
Dark-colored soy sauce F	Super	Soybean, wheat, salt	3.6
Light-colored soy sauce A	Ordinary	Salt, soybean, wheat, ethanol	3.7
Light-colored soy sauce B	Ordinary	Soybean, wheat, rice, salt, ethanol	3.4
White soy sauce	Ordinary	Wheat, soybean, salt, ethanol	1.5

Table 7.22 The contents of γ-Glu-Val-Gly in various fermented shrimp paste condiments

Sample	Country of origin	Contents of γ-Glu-Val-Gly* (mg/kg)
Terasi	Indonesia	5.2 ± 0.3
Bagoong	Philippines	1.0 ± 0.1
Xia jiang	China	0.9 ± 0.1

*mean ± standard deviation ($n = 3$)

summarized in Table 7.24. The average content of γ-Glu-Val-Gly in fish sauces was 4.4 mg/L, that in soy sauces was 4.0 mg/L, that in fermented shrimp paste was 2.4 mg/L, and that in beer was 0.12 mg/L. These results indicate that γ-Glu-Val-Gly is widely distributed in fermented foods such as fish sauces, soy sauces, fermented shrimp paste, and beer. In addition, it was revealed that the contents of γ-Glu-Val-Gly in fermented seasonings such as fish sauce, soy sauce, and fermented shrimp paste were higher than those of brew such as beer. The variation of the contents would be because of the difference in the raw materials, fermentation condition, and so on. However, further investigation should be performed to clarify the reasons behind the variation in γ-Glu-Val-Gly contents.

7.7.4.2 Correlation Between γ-Glu-Val-Gly Content and Quality Grade

To clarify the contribution of γ-Glu-Val-Gly to the sensory quality of various fermented foods, the correlation between the content of the peptide and index of the quality of soy sauce (Yokotsuka 1986) was investigated. As shown in

Table 7.23 The contents of γ-Glu-Val-Gly in various alcoholic beverages

Samples	Country of origin	Contents of γ-Glu-Val-Gly (mg/L)[a]
White wine A	Japan	n.d.[b]
White wine B	France	n.d.
Red wine A	Japan	n.d.
Red wine B	France	n.d.
Rice wine (*sake*) A	Japan	n.d.
Rice wine (*sake*) B	Japan	n.d.
Rice wine (*sake*) C	Japan	n.d.
Rice wine (*sake*) D	Japan	n.d.
Beer A	UK	0.13 ± 0.009
Beer B	Italy	0.18 ± 0.011
Beer C	Japan	0.10 ± 0.006
Beer D	France	0.09 ± 0.004
Beer E	Spain	0.15 ± 0.008
Beer F	Germany	0.09 ± 0.011
Beer G	Belgium	0.11 ± 0.004
Beer H	Germany	0.08 ± 0.007

Limit of detection (in wine and rice wine) = 0.03 mg/L
Limit of detection (in beer) = 0.01 mg/L
[a]Mean ± standard deviation ($n = 4$)
[b]*n.d.* not detected

Table 7.24 The contents of γ-Glu-Val-Gly in various fermented foods

Samples	Sample size (n)	Contents of γ-Glu-Val-Gly (mg/L)		
		Average	Minimum	Max
Fish sauces				
Total	17	4.4	<LOQ	12.6
Nampra (Thailand)	5	2.3	1.2	2.7
Nuoc mum (Vietnam)	5	11.6	10.4	12.6
Yu-lu (China)	2	0.6	<LOQ	1.1
Japanese fish sauce	4	0.8	<LOQ	2.8
Garum (Italy)	1	0.4	0.4	0.4
Soy sauces (Japan)				
Total	9	4.0	1.5	5.3
Dark-colored soy sauce	6	4.6	3.1	5.3
Light-colored soy sauce	2	3.6	3.4	3.7
White soy sauce	1	1.5	1.5	1.5
Fermented shrimp paste				
Total	3	2.4	0.9	5.2
Terasi (Indonesia)	1	5.2	5.2	5.2
Bagoong (Philippines)	1	1.0	1.0	1.0
Xia jiang (China)	1	0.9	0.9	0.9
Beer				
Total	8	0.12	0.08	0.15
Top-fermented beer	2	0.11	0.09	0.11
Bottom-fermented beer	6	0.12	0.08	0.15

LOQ limit of quantification

Table 7.25 Standard for Japanese soy sauces

Type of soy sauce	Class	Grade	Contents (g/dL)	
			Total nitrogen	Dry matter of extract
Dark-colored	Standard		>1.20	–
	Upper		>1.35	>14
	Special	Ordinary	>1.50	>16
		Super	>1.65	>16
		Ultrasuper	>1.80	>16
Light-colored	Standard		>0.95	–
	Upper		>1.05	>12
	Special	Ordinary	>1.15	>14
		Super	>1.15	>15.4
		Ultrasuper	>1.15	>16.8
Light-colored	Standard		<0.90	>10
	Upper		<0.90	>13
	Special	Ordinary	<0.80	>16
		Super	<0.80	>17.6
		Ultrasuper	<0.80	>19.2

Fig. 7.8 Correlation between the contents of γ-Glu-Val-Gly and the quality of dark-colored soy sauces. Each bar indicates the mean value, and plots indicate the contents in each sample

Table 7.25, Japanese dark-colored soy sauces (*koikuchi-shoyu*) are divided into three classes according to the total nitrogen content, special, higher, and standard (Yokotsuka 1961). In addition, the soy sauces in the special class are divided into three grades, ultrasuper, super, and ordinary. All of the six dark-colored soy sauces tested in this study are of special class, and they consist of one ultrasuper grade, three super grades, and two ordinary grades. The γ-Glu-Val-Gly content in the ultrasuper grade was highest (6.1 mg/L), followed by the super grade (average content = 4.7 mg/L), and the ordinary grade (average content = 3.7 mg/L). The Spearman's rank correlation test indicated that the content of γ-Glu-Val-Gly was positively correlated with the quality grade of commercial soy sauce (correlation coefficient $\rho = 0.810$, $p < 0.05$) (Fig. 7.8). These results suggest that γ-Glu-Val-Gly contributed to the sensory quality of fermented foods.

In our laboratory, the contents of γ-Glu-Val-Gly in other fermented foods and the contribution of this peptide to the quality of foods is now in progress.

7.7.5 Conclusion

The content of a *kokumi* peptide, γ-Glu-Val-Gly, in various foods was analyzed by LC/MS/MS followed by derivatization with AQC. The quantitative analysis indicated that γ-Glu-Val-Gly is distributed in various fermented foods, such as fish sauce, soy sauce, fermented shrimp paste, and beer. Furthermore, the content of this peptide has been positively correlated with the quality grade of soy sauces, suggesting that γ-Glu-Val-Gly contributes to the sensory quality of fermented foods.

7.8 General Discussion

In this chapter, the effect of a *kokumi* peptide, γ-glutamyl-valyl-glycine (γ-Glu-Val-Gly), on the sensory characteristics of various foods was investigated.

In Sect. 7.1 (background) and Sect. 7.7, distribution of *kokumi* γ-glutamyl peptides in foods and distribution of γ-Glu-Val-Gly in various fermented foods are described. In addition, the results in Sect. 7.7 showed that the content of γ-Glu-Val-Gly is positively correlated with the quality grade of soy sauces, suggesting that γ-Glu-Val-Gly contributes to the sensory quality of fermented foods.

In Sect. 7.2, glutathione (GSH), a representative *kokumi* peptide, and γ-Glu-Val-Gly were added to chicken soup prepared using a commercial chicken consommé cube, and the sensory effects were compared. As shown in Fig. 7.2, both GSH and γ-Glu-Val-Gly increased the intensity of "continuity" (taste intensity at 20 seconds after tasting), "mouthfulness" (the reinforcement of taste sensation throughout the mouth not just on the tongue), and "thickness" (taste intensity at 5 s after tasting). This result suggests that γ-Glu-Val-Gly also has sensory characteristics as a *kokumi* substance. From the results that the effects of 20 ppm (66.0 μM) γ-Glu-Val-Gly were almost same levels as those of 200 ppm (651 μM) GSH, γ-Glu-Val-Gly was considered to be a potent *kokumi* substance.

In Sect. 7.3, the effect of γ-Glu-Val-Gly on the sensory characteristics of handmade chicken consommé was investigated. The evaluation was performed by a descriptive method. The sensory attributes were selected, and the definition and reference sample for each attribute were determined via discussion by panelists. As a result, 17 attributes, in which 9 attributes for flavor and 8 attributes for other mouthfeel, were selected. Especially, besides the attributes on the chicken, because the coating sensation was well recognized when the panelists evaluated the chicken consommé with γ-Glu-Val-Gly during the project-specific panel orientation session, the attributes "mouth-coating" and "tongue-coating" were added to the list. "Mouth-coating" was defined as the degree to which there is a leftover residue, a slick, powdery, or fatty coating or film on the mouth that is difficult to clear. "Tongue-coating" was defined as the degree to which there is a leftover residues, a slick, powdery, or fatty coating or film on the tongue that is difficult to clear. Overall, the panelists defined the 17 sensory attributes for chicken consommé listed in Table 7.2:

nine taste and flavor attributes (total flavor, total chicken/meaty flavor, chicken flavor, bone and marrow flavor, roasted flavor, total vegetable flavor, richness, salty, and umami), seven texture/mouthfeel attributes (viscosity, mouthfulness, mouth-coating, tongue-coating, salivating, total trigeminal, and swelling perception of soft tissue), and one aftertaste (total aftertaste). Definitions of these sensory attributes and the references are shown in Table 7.6. The additional effect of 5 ppm (16.5 μM) γ-Glu-Val-Gly on chicken consommé is revealed in Table 7.7 and Fig. 7.3. γ-Glu-Val-Gly significantly increased the intensities of umami, mouthfulness, and mouth-coating, and it tended to increase the intensity of viscosity. A recent study has suggested that *kokumi* peptides such as GSH and γ-Glu-Val-Gly enhance the intensity of umami if they are added to 0.5% MSG solution (Ohsu et al. 2010), an observation consistent with the present study. Additionally, in the descriptive analysis, umami has been defined as not only the "taste of MSG" but also as "the mouth-filling sensation of compounds such as glutamates that is savory, brothy, meaty, rich, full, and complex, which is common to many foods such as soy sauce, stocks, ripened cheese, shellfish, mushrooms, ripened tomatoes, cashews, and asparagus." Therefore, it appears that the enhancement of umami in chicken consommé includes the enhancement of sensations such as richness and complexity The present results also suggest that γ-Glu-Val-Gly also enhanced mouthfulness. A study in Sect. 7.2 demonstrated that adding γ-Glu-Val-Gly at 20 ppm to chicken soup prepared from commercial chicken consommé powder significantly enhanced mouthfulness which is consistent with the present study (Ohsu et al. 2010). In Sect. 7.2, mouthfulness was defined as the reinforcement of the taste sensation throughout the mouth and not just on the tongue, while it was defined as the perception that the sample fills the whole mouth, is blooming, or is growing, a full-bodied sensation when the sample is held in the mouth in Sect. 7.3. These expressions were similar, suggesting that both of the panelists in the study of Sect. 7.2 (trained Japanese assessors) and those in Sect. 7.3 (trained panelists who are residents in the USA) could detect the increase of mouthfulness when γ-Glu-Val-Gly was added to chicken soup. The additional levels were different in these two studies, and it is considered because of the difference in the composition of chicken soups. Regarding other γ-glutamyl peptides, it has been reported that several *kokumi* γ-glutamyl peptides enhanced mouthfulness in food systems. Ueda et al. (1997) reported that the addition of GSH (γ-Glu-Cys-Gly) enhanced the intensity of mouthfulness in model beef meat extract. In addition, Ohsu et al. (2010) also reported that the addition of GSH enhanced the intensity of mouthfulness in chicken soup. Furthermore, it has been reported that γ-glutamyl peptides such as γ-Glu-Val, γ-Glu-Leu, and γ-Glu-Cys-β-Ala found as *kokumi*-active peptides in edible beans enhanced mouthfulness when they were added to chicken broth (Dunkel et al., 2007). In addition, it has been reported that γ-Glu-Glu, γ-Glu-Gly, γ-Glu-His, γ-Glu-Gln, γ-Glu-Met, and γ-Glu-Leu were the key components which impart long-lasting mouthfulness of matured Gouda cheese. From these observations, it is demonstrated that many *kokumi* γ-glutamyl peptides enhance the intensity of mouthfulness when added to chicken soup. In addition, the results in Sects. 7.2 and 7.3 indicated that γ-Glu-Val-Gly also increased the intensity of mouthfulness when added to chicken soup. Furthermore, the addition of γ-Glu-Val-Gly at 5 ppm significantly increased the intensity of mouth-coating. It has been generally known that mouth-coating sensation is evoked by the addition of hydrocolloids such as xanthan gum and locust bean gum, carrageenan (Arocas et al. 2010; Flett et al. 2010), and fat-containing food materials such as dairy fat emulsion (Flett et al. 2010). Despite these observations, there have been no reports of a peptide which exhibited the mouth-coating sensation. Therefore, this is the first report which has demonstrated the mouth-coating effect of peptides. In addition, the addition of this peptide tended to increase the intensity of viscosity. Although the viscosity of consommé did not change significantly by adding 5 ppm of γ-Glu-Val-Gly (data not shown), an enhancement of the mouth-coating sensation and trend to increase of viscosity was observed. The mechanism of this enhancement is interesting and should be clarified by further investigations.

The effect of γ-Glu-Val-Gly on the reduced-fat peanut butter was investigated with 19 trained Chinese panelists. In this study 40 ppm (132 μM) of γ-Glu-Val-Gly was added to a reduced-fat peanut butter and evaluated by descriptive analysis. Five attributes, peanut flavor, thickness of taste, aftertaste, continuity, and oiliness, were used for the sensory evaluation. Peanut flavor was defined as the flavor intensity reminiscent of peanut. Thickness of taste was defined as the enhancement of taste intensity with maintaining the balance of taste. Aftertaste was defined as the total taste intensity at 5 sec after tasting, and continuity was defined as the total taste intensity at 20 sec after tasting. Oiliness was defined as the oily mouth-coating sensation. As shown in Table 7.12 and Fig. 7.4, the addition of 40 ppm (132 μM) γ-Glu-Val-Gly significantly enhanced the intensities of thickness of taste, aftertaste, and oiliness. Similar to the results on the effect of this peptide on chicken soup prepared from commercial chicken consommé (Sect. 7.2), γ-Glu-Val-Gly increased the intensity of thickness of taste, suggesting that this peptide can enhance the thickness of taste both in savory and sweet foods. Interestingly, the addition of γ-Glu-Val-Gly increased the intensity of oiliness, which is defined as the oily mouth-coating sensation in reduced-fat peanut butter, although the viscosity of peanut butter did not change significantly by adding 40 ppm of γ-Glu-Val-Gly (data not shown). As mentioned above, this peptide increased the intensity of mouth-coating sensation in handmade chicken consommé (Sect. 7.3). Therefore, it was considered that γ-Glu-Val-Gly can enhance a mouth-coating sensation both in chicken consommé and reduced-fat peanut butter. In the evaluation of the reduced-fat peanut butter, γ-Glu-Val-Gly was added at the concentration of 40 ppm (132 μM), which was selected as an appropriate concentration by preliminary test. The additional level was higher than that in handmade chicken consommé (5 ppm; 16.5 μM). Although the detailed reason is not clarified, the difference of the appropriate concentration is because of the difference of the composition of foods.

The effect of γ-Glu-Val-Gly on the reduced-fat custard cream was investigated with 19 trained Chinese panelists. In this study, 40 ppm (132 μM) of γ-Glu-Val-Gly was added to a reduced-fat custard cream and evaluated by descriptive analysis. Five attributes, thickness of taste, continuity, aftertaste, vanilla flavor, and cream flavor, were used for the sensory evaluation. Thickness of taste was defined as the enhancement of taste intensity with maintaining the balance of taste. Continuity was defined as the total taste intensity at 20 sec after tasting, and aftertaste was defined as the total taste intensity at 5 sec after tasting. Vanilla flavor was defined as the flavor intensity reminiscent of vanilla. Cream flavor was defined as the flavor intensity reminiscent of cream. As shown in Table 7.16 and Fig. 7.5, the addition of 40 ppm (132 μM) γ-Glu-Val-Gly significantly enhanced the intensities of thickness of taste and continuity of taste and tended to increase the intensity of aftertaste. Similar to the results on the effect of this peptide on chicken soup prepared from commercial chicken consommé (Sect. 7.2), γ-Glu-Val-Gly increased the intensity of thickness of taste, suggesting that this peptide can enhance the thickness of taste both in savory and sweet foods. In addition, similar to the results on the effect of this peptide on chicken soup prepared from commercial chicken consommé (Sect. 7.2), γ-Glu-Val-Gly increased the intensity of continuity of taste, suggesting that this peptide can enhance the continuity of taste both in savory and sweet foods. In the evaluation of the reduced-fat custard cream, γ-Glu-Val-Gly was added at the concentration of 40 ppm (132 μM), which was selected as an appropriate concentration by preliminary test. The additional level was higher than that in handmade chicken consommé (5 ppm; 16.5 μM). Although the detailed reason is not clarified, the difference of the appropriate concentration is because of the difference of the composition of foods.

The effect of γ-Glu-Val-Gly on the reduced-fat French dressing was investigated with 19 trained Chinese panelists. In this study, 40 ppm (132 μM) of γ-Glu-Val-Gly was added to a reduced-fat French dressing and evaluated by descriptive analysis. Five attributes, spice flavor, thickness of taste, initial taste, aftertaste, and smoothness were used for the sensory evaluation. Spice flavor was defined as flavor intensity reminiscent of spice, especially paprika and garlic. Thickness of taste was defined as the enhancement of taste intensity with maintaining the balance of taste. Initial taste was defined as the total taste intensity at 1 s after tasting. Aftertaste was defined as the total taste intensity at 5 s after tasting. Smoothness was defined as the oily smooth sensation. As shown in Table 7.20 and Fig. 7.6, the addition of 40 ppm (132 μM) γ-Glu-Val-Gly significantly enhanced the intensities of aftertaste and tended to increase the intensity of initial taste. Similar to the results on the effect of this peptide on chicken soup prepared from commercial chicken consommé (Sect. 7.2), γ-Glu-Val-Gly increased the intensity of aftertaste, suggesting that this peptide can enhance the aftertaste in reduced-fat French dressing. In the evaluation of the reduced-fat French dressing, γ-Glu-Val-Gly was added at the concentration of 40 ppm (132 μM), which was selected as an appropriate concentration by preliminary test. The additional level was higher than that in handmade chicken consommé (5 ppm; 16.5 μM). Although the detailed reason is not clarified, this difference in the required concentration may be due to the difference in the composition of these foods.

The effect of γ-Glu-Val-Gly on foods such as handmade chicken consommé, reduced-fat peanut butter, reduced-fat custard cream, and reduced-fat French dressing was evaluated by a descriptive analysis. The addition of this peptide enhanced umami, mouthfulness, and mouth-coating in chicken consommé; enhanced thickness of taste, aftertaste, and oiliness (oily mouth-coating sensation) in reduced-fat peanut butter; enhanced thickness of taste in reduced-fat custard cream; and enhanced aftertaste in reduced-fat French dressing. These results suggest that this peptide can improve the flavor of both savory and sweet food. Yamaguchi and Kimizuka (1979) reported that the addition of monosodium glutamate (MSG) to beef consommé or chicken soups (chicken consommé, chicken noodle soup, and cream of chicken soup) enhanced continuity, mouthfulness, and thickness. These results suggest that MSG, an umami substance, increases not only the intensity of umami (taste of MSG aqueous solution) but also that of continuity, mouthfulness, and thickness, *koku*-related attributes. Therefore, MSG can be one of *koku*-imparting substances. Yamaguchi and Kimizuka (1979) also reported that the addition of sugar to caramel custard or Bavarian cream increased the intensity of continuity, mouthfulness, and thickness, suggesting that sugar (sucrose), a sweetener, increases not only the intensity of sweetness but also that of continuity, mouthfulness, and thickness, *koku*-related attributes. Therefore, sugar (sucrose) can be one of *koku*-imparting substances. However, because MSG and sugar have tastes themselves, MSG is usually used in savory foods and sugar is usually used in sweet foods. On the other hands, γ-Glu-Val-Gly, a *kokumi* substance, enhanced *koku*-related attributes such as mouthfulness and thickness of taste both in savory and sweet food because it has no taste itself at the concentrations used in sensory evaluation. This is a unique character of *kokumi* substance.

The study in this chapter demonstrated a possibility that the use of a *kokumi* peptide, γ-Glu-Val-Gly, improves the flavor of various foods including savory and sweet foods. Further potential of this peptide should be investigated with studies on the perceptive mechanism of *kokumi* substances.

Acknowledgments We sincerely thank Dr. Kiyoshi Miwa, Dr. Tohru Kouda, and Dr. Yuzuru Eto of Ajinomoto Co., Inc., for their encouragement and continued support of this work. We thank Ms. Sharon McEvoy and Dr. Dawn Chapman of the National Food Laboratory LLC for their cooperation and a valuable discussion. We thank Ms. Yang Xiao Qing and Ms. Tao Wei in Shanghai Ajinomoto R&D Center for their cooperation. We sincerely thank Dr. Hiroshi Miyano, Dr. Toshimi Kizukoshi, Ms. Naoko Kageyama, Ms. Yumiko Kato, Ms. Junko Yamazaki, and Ms. Yuko Kai. We are grateful to Dr. Chinatsu Kasamatsu, Dr. Hiroya Kawasaki, Mr. Hiroaki Nagasaki, Mr. Tomohiko Yamanaka, Mr. Fusataka Kenmotsu, Mr. Toshifumi Imada, Mr. Takashi Miyaki, Mr. Takaho Tajima, Mr. Shuichi Jo, Mr. Keita Sasaki, and Ms. Takako Hirose of Ajinomoto Co., Inc., for their assistance. We are grateful to the panelists who have participated on the sensory evaluation.

References

Abdel-Haleem AM, Awad RA (2015) Some quality attributes of low fat ice cream substituted with hulless barley flour and barley β-glucan. J Food Sci Technol 52:6425–6434

Aghdaei SSA, Aslami M, Geefan SB, Ranjbar A (2014) Application of Isfarzeh seed (Plantago ovate L.) mucilage as a fat mimetics in mayonnaise. J Food Sci Technol 51:2748–2754

Amelia I, Drake MA, Nelson B, Barbano DM (2013) A new method for the production of low-fat Cheddar cheese. J Dairy Sci 96:4870–4884

Armenta JM, Cortes DF, Pisciotta JM, Shuman JL, Blakeslee K, Rasoloson D (2010) Sensitive and rapid method for amino acid quantification in Malaria biological samples using AccQ-Tag ultra performance liquid chromatography-electrospray ionization-MS/MS with multiple reaction monitoring. Anal Chem 82:548–558

Arocas A, Sanz T, Varela P, Fiszman SM (2010) Sensory properties determined by starch type in white sauces: effects of freeze/thaw and hydrocolloid addition. J Food Sci 75:S132–S140

Azari-Anpar M, Khomeiri M, Ghafouri-Oskuei H, Aghajani N (2017) Response surface optimization of low-fat ice cream production by using resistant starch and maltodextrin as a fat replacing agent. J Food Sci Technol 54:1175–1183

Bernardino-Nicanor A, Hinojosa-Hernandez EN, Juarez-Goiz JMS, Montanez-Soto JL, Ramirez-Ortiz ME, Gonzalez-Cruz L (2015) Quality of Opuntia robusta and its use in development of mayonnaise like product. J Food Sci Technol 52:343–350

Dawid C, Hofmann T (2012) Identification of sensory-active phytochemicals in asparagus (Asparagus officinalis L.). J Agric Food Chem 60:11877–11888

de Souza Fernandes D, Leonel M, Del Bem MS, Mischan MM, Garcia ÉL, Dos Santos TPR (2017) Cassava derivatives in ice cream formulations: effects on physicochemical, physical and sensory properties. J Food Sci Technol 54:1357–1367

Dunkel A, Koster J, Hofmann T (2007) Molecular and sensory characterization of γ-glutamyl peptides as key contributors to the kokumi taste of edible beans (Phaseolus vulgaris L.). J Agric Food Chem 55:6712–6719

Fernandes SS, Mellado MMS (2018) Development of mayonnaise with substitution of oil or egg yolk by the addition of chia (*Salvia Hispanica* L.) mucilage. J Food Sci 83:74–83

Flett KL, Duizer LM, Goff D (2010) Perceived creaminess and viscosity of aggregated particles of casein micelles and κ-carrageenan. J Food Sci 75:S255–S261

Furukawa H (1977) The sensory test and the selection of panellists. In: Proc. 7th Symp. Sensory inspection, pp 111

Guo Y, Zhang X, Hao W, Xie Y, Chen L, Li Z, Zhu B, Feng X (2018) Nano-bacterial cellulose/soy protein isolate complex gel as fat substitutes in ice cream model. Carbohydr Polym 198:620–630

Hillmann H, Behr J, Ehrmann MA, Vogel RF, Hofmann T (2016) Formation of kokumi-enhancing γ-glutamyl dipeptides in parmesan cheese by means of γ-glutamyltransferase activity and stable isotope double-labelling studies. J Agric Food Chem 64:1784–1793

Kuroda M, Nagaba N, Tsubuku T, Kawajiri H, Ueda Y (1997) Simultaneous determination of glutathione, gamma-glutamylcysteine and cysteine in commercial yeast extract by HPLC with fluorimetric detection. Food Sci Tech Int, 3, 239–240

Kuroda M, Kato Y, Yamazaki J, Kageyama N, Mizukoshi T, Miyano H, Eto Y (2012a) Determination of γ-glutamyl-valyl-glycine in raw scallop and processed scallop products using high performance liquid chromatography-tandem mass spectrometry. Food Chem 134:1640–1644

Kuroda M, Kato Y, Yamazaki J, Kai Y, Mizukoshi T, Miyano H, Eto Y (2012b) Determination and quantification of γ-glutamyl-valyl-glycine in commercial fish sauces. J Agric Food Chem 60:7291–7296

Kuroda M, Kato Y, Yamazaki J, Kai Y, Mizukoshi T, Miyano H, Eto Y (2013) Determination and quantification of the kokumi peptide, γ-glutamyl-valyl-glycine, in commercial soy sauces. Food Chem 141:823–828

Kuroda M, Miyamura N, Mizukoshi T, Miyano H, Kouda T, Eto. Distribution of a kokumi peptide, γ-Glu-Val-Gly, in various fermented foods and the possibility of its contribution to the sensory quality of fermented foods. Ferment Tech 2015, 4:2, https://doi.org/10.4172/2167-7972.1000121

Liou BK, Grun IU (2007) Effect of fat level on the perception of five flavor chemicals in ice cream with or without fat mimetics by using a descriptive test. J Food Sci 72:S595–S604

Liu H, Xu XM, Guo SD (2007) Rheological, texture and sensory properties of low-fat mayonnaise with different fat mimetics. LWT Food Sci Technol 40:946–954

Maruyama Y, Yasuda R, Kuroda M, Eto Y (2012) *Kokumi* substances, enhancers of basic tastes, induce responses in calcium-sensing receptor expressing taste cells. PLoS ONE, 7, e34489

McClements DJ, Demetriades K (1998) An integrated approach to the development of reduced-fat food emulsions. Crit Rev Food Sci Nutr 38:511–536

Mellies MJ, Vitale C, Jandacek RJ, Lamkin GE, Glueck CJ (1985) The substitution of sucrose polyester for dietary fat in obese, hypercholesterolemic outpatients. Am J Clin Nutr 41:1–12

Miyamura N, Kuroda M, Kato Y, Yamazaki J, Mizukoshi T, Miyano H, Eto Y (2014) Determination and quantification of a kokumi peptide, γ-glutamyl-valyl-glycine, in fermented shrimp paste condiments. Food Sci Tech Res 20:699–703

Miyamura N, Iida Y, Kuroda M, Kato Y, Yamazaki J, Mizukoshi T, Miyano H (2015a) Determination and quantification of kokumi peptide, γ-glutamyl-valyl-glycine, in brewed alcoholic beverages. J Biosci Bioeng 120:311–314

Miyamura N, Jo S, Kuroda M, Kouda T (2015b) Flavour improvement of reduced-fat peanut butter by addition of a kokumi peptide, γ-glutamyl-valyl-glycine. Flavour 4:16

Miyamura N, Kuroda M, Kato Y, Yamazaki J, Mizukoshi T, Miyano H (2016) Quantitative analysis of γ-glutamyl-valyl-glycine in fish sauces fermented with koji by LC/MS/MS. Chromatography 37:39–42

Ohsu T, Amino Y, Nagasaki H, Yamanaka T, Takeshita S, Hatanaka T, Maruyama Y, Miyamura N, Eto Y (2010) Involvement of the calcium-sensing receptor in human taste perception. J Biol Chem 285:1016–1022

Patel AS, Jana AH, Aparnathi KD, Pinto SV (2010) Evaluating sago as a functional ingredient in dietetic mango ice cream. J Food Sci Technol 47:582–585

Pimentel TC, Cruz AG, Prudencio SH (2013) Influence of long-chain inulin and lactobacillus paracasei subspecies paracasei on the sensory profile and acceptance of a traditional yogurt. J Dairy Sci 96:6233–6624

Roman L, Martinez MM, Gomez M (2015) Assessing of the potential of extruded flour paste as fat replacer in O/W emulsion: a rheological and microstructural study. Food Res Int 74:72–79

Scharbert S, Holzmann N, Hofmann T (2004) Identification of the astringent taste compounds in black tea infusions by combining instrumental analysis and human bioresponse. J Agric Food Chem 52:3498–3508

Schwarz B, Hofmann T (2009) Identification of novel orosensory active molecules in cured vanilla beans (Vanilla planifolia). J Agric Food Chem 57:3729–3737

Sharma M, Singh AK, Yadav DN (2017) Rheological properties of reduced fat ice cream mix containing octenyl succinylated pearl millet starch. J Food Sci Technol 54:1638–1645

Shibata M, Hirotsuka M, Mizutani Y, Takahashi H, Kawada T, Matsumiya K, Hayashi Y, Matsumura Y (2017) Isolation and characterization of key contributors to the "kokumi" taste in soybean seeds. Biosci Biotechnol Biochem 81:2168–2177

Su H-P, Lien C-P, Lee T-A, Ho J-H (2010) Development of low-fat mayonnaise containing polysaccharide gums as functional ingredients. J Sci Food Agric 90:806–812

Sun C, Liu R, Liang B, Wu T, Sui W, Zhang M (2018) Microparticulated whey protein –pectin complex: a texture-controllable gel for low-fat mayonnaise. Food Res Int 108:151–160

Toelstede S, Dunkel A, Hofmann T (2009) A series of kokumi peptides impart the long-lasting mouthfulness of matured Gouda cheese. J Agric Food Chem 57:1440–1448

Tomaschunas M, Zorb R, Fischer J, Kohn E, Hinrichs J, Busch-Stockfisch M (2013) Changes in sensory properties and consumer acceptance of reduced fat pork Lyon-style and liver sausages containing inulin and citrus fiber as fat replacers. Meat Sci 95:629–640

Ueda Y, Sakaguchi M, Hirayama K, Miyajima R, Kimizuka A (1990) Characteristic flavor constituents in water extract of garlic. Agric Biol Chem 54:163–169

Ueda Y, Tsubuku T, Miyajima R (1994) Composition of sulfur-containing components in onion and their flavor characters. Biosci Biotechnol Biochem 61:108–110

Ueda Y, Yonemitsu M, Tsubuku T, Sakaguchi M, Miyajima R (1997) Flavor characteristics of glutathione in raw and cooked foodstuffs. Biosci Biotechnol Biochem 61:1977–1980

Wendin K, Aaby K, Edris A, Ellekjaer MR, Albin R, Bergensahl B, Johansson L, Willers EP, Solheim R (1997) Low-fat mayonnaise: influences of fat content, aroma compounds and thickeners. Food Hydrocol 11:87089

Yamaguchi S, Kimizuka A (1979) Psychometric studies on the taste of monosodium glutamate. In: Filer LJ Jr, Garattini S, Kare MR, Reynolds WA, Wurtman RJ (eds) Glutamic acid: advances in biochemistry and physiology. Raven Press, New York, pp 35–54

Yildrim M, Sumnu G, Sahin S (2016) Rheology, particle-size distribution, and stability of low-fat mayonnaise produced via double emulsions. Food Sci Biotechnol 25:1613–1618

Yokotsuka T (1961) Aroma and flavor of Japanese soy sauce. Adv Food Res 10:75–134

Yokotsuka T (1986) Soy sauce biochemistry. Adv Food Res 30:195–329

Zhang H, Chen J, Li J, Wei C, Ye X, Shi J, Chen S (2018) Pectin from citrus canning wastewater as potential fat replacer in ice cream. Molecules 23(4). pii: E925. https://doi.org/10.3390/molecules23040925

Chapter 8
Mechanism of *Kokumi* Substance Perception: Role of Calcium-Sensing Receptor (CaSR) in Perceiving *Kokumi* Substances

Yutaka Maruyama and Motonaka Kuroda

Keywords *Kokumi* substance · Calcium-sensing receptor · CaSR · Glutathione · γ-Glu-Val-Gly · γ-Glutamyl-valyl-glycine · γ-Glu-Nva-Gly · γ-Glutamyl-norvalyl-glycine · γ-Glu-Nva · γ-Glutamyl-norvaline

8.1 Involvement of the Calcium-Sensing Receptor (CaSR) in Perception of *Kokumi* Substances

Abstract Some foods are known to have flavor sensations, such as continuity, mouthfulness, and thick flavor, which cannot be explained by the five basic tastes alone. It has been demonstrated that these sensations are evoked by the addition of *kokumi* substances, which are flavor modifiers with no actual taste themselves. However, the mechanism is poorly understood. Investigation of the perception of amino acids and peptides revealed that γ-glutamyl-cysteinyl-glycine (GSH) was an agonist of calcium-sensing receptor (CaSR). Therefore, we hypothesized that CaSR was involved in perception of *kokumi* substances. In fact, all of the CaSR-activating substances we tested were effective as *kokumi* substances, and there was a positive correlation between CaSR activity and *kokumi* intensity of γ-glutamyl peptides. Furthermore, *kokumi* intensities of GSH and γ-Glu-Val-Gly, a potent *kokumi* peptide, were significantly reduced by a CaSR inhibitor, NPS-2143. These results suggest that CaSR is involved in the perception of *kokumi* substances.

Y. Maruyama
Institute for Innovation, Ajinomoto Co., Inc., Kawasaki, Kanagawa, Japan

M. Kuroda (✉)
Institute of Food Sciences & Technologies, Ajinomoto,
Suzuki-cho, Kawasaki, Kanagawa, Japan
e-mail: motonaka_kuroda@ajinomoto.com

T. Nishimura, M. Kuroda (eds.), *Koku in Food Science and Physiology*,
https://doi.org/10.1007/978-981-13-8453-0_8

8.1.1 Introduction

Taste and aroma are important factors in determining the flavor of foods. The five basic tastes are sweet, salty, sour, bitter, and umami with each taste being recognized by specific receptors and associated with particular nerve pathways. However, foods have sensory attributes that cannot be explained by aroma and the five basic tastes alone, and these attributes include texture, continuity, complexity, and mouthfulness. Ueda et al. (1990) investigated the flavoring effect of a diluted garlic extract that enhanced continuity, mouthfulness, and thickness when added to an umami solution and attempted to isolate and identify the key compounds responsible for this effect. They found that sulfur-containing compounds were involved in this flavoring effect, including S-allyl-cysteine sulfoxide (alliin), S-methyl-cysteine sulfoxide, γ-glutamyl-allyl-cysteine, and glutathione (γ-glutamyl-cysteinyl-glycine; GSH). They also investigated the flavoring effect of onion extract and identified S-propenyl-L-cysteine sulfoxide, alliin, and γ-glutamyl-S-propenyl-L-cysteine sulfoxide as major components that enhanced continuity, mouthfulness, and thickness when added to an umami solution (Ueda et al. 1994). These compounds only have a minimal flavor when added to water, but substantially enhance thickness, continuity, and mouthfulness if added to an umami solution or to other foods (Ueda et al. 1997). Ueda and colleagues proposed that substances with these properties should be referred to as *kokumi* substances.

Dunkel et al. subsequently investigated *kokumi* substances in edible beans, and they identified γ-glutamyl-leucine, γ-glutamyl-valine, and γ-glutamyl-cysteinyl-β-alanine as key *kokumi* substances (Dunkel et al. 2007). Using a similar approach, several γ-glutamyl peptides such as γ-glutamyl-glutamic acid, γ-glutamyl-glycine, γ-glutamyl-glutamine, γ-glutamyl-methionine, γ-glutamyl-leucine, and γ-glutamyl-histidine have been identified as *kokumi* substances in matured Gouda cheese (Toelstede and Hofmann 2009). These studies based on fractionation and sensory evaluation have revealed that various foods contain *kokumi* substances. However, the molecular mechanisms underlying perception of such substances have not been clarified.

The extracellular calcium-sensing receptor, CaSR, is a classic seven-transmembrane-spanning, G protein-coupled receptor (GPCR) belonging to family C of the GPCR superfamily (Brown et al. 1993). This receptor has been identified in several tissues and organs, including the parathyroid gland and kidney. It plays a central role in extracellular calcium homeostasis in mammals (Chattopadhyay et al. 1997). An increase in the blood calcium level is sensed by CaSR, which suppresses parathyroid hormone secretion, stimulates calcitonin secretion, and induces urinary calcium excretion to reduce the blood calcium level to the normal range. It has been shown that CaSR is not only expressed in the parathyroid glands and kidney, but also in many other organs, including the liver, heart, lung, gastrointestinal tract, pancreas, and central nervous system, suggesting that it is involved in a wide range of biological functions (Brown and MacLeod 2001). It has been reported that CaSR is activated by various substances, including cations such as Ca^{2+}, Mg^{2+}, and Gd^{3+}, basic peptides such as protamine and polylysine, and polyamines such as spermine (Brown and MacLeod 2001).

CaSR is expressed in a subpopulation of taste cells in mice and rats (Bystrova et al. 2010; San Gabriel et al. 2009), suggesting a potential role in taste cell biology. Ninomiya and colleagues reported that a group of gustatory afferent nerve fibers respond to calcium and magnesium in mice (Ninomiya et al. 1982). Tordoff and coworkers investigated taste perception of calcium and the physiological mechanisms underlying calcium intake, appetite, and homeostasis, and they demonstrated that calcium deprivation increases the palatability of calcium (McCaughey et al. 2005). These findings suggest the existence of a calcium transduction mechanism in taste cells. However, the physiological role of CaSR agonists other than calcium is not clear. Recently, Bystrova et al. reported that CaSR is expressed in a subset of taste cells and that CaSR agonists (NPS R-568, neomycin, and several L-amino acids) induced a response from isolated taste cells (Bystrova et al. 2010).

Based on the above reports, we investigated CaSR in the present study to clarify mechanism involved in the perception of *kokumi* substances.

8.1.2 Materials and Methods

8.1.2.1 Materials

CaSR agonists used in human sensory evaluations were of food additive grade and were purchased from the manufacturer as reported previously (Ohsu et al. 2010). Cinacalcet (Rodriguez et al. 2005) and NPS-2143 were chemically synthesized by methods described in the literature. γ-Glutamyl peptides such as γ-Glu-Cys-Gly, γ-Glu-Cys, γ-Glu-Ala, γ-Glu-Abu-Gly, and γ-Glu-Val were purchased from the manufacturers, as reported previously (Ohsu et al. 2010). Other γ-glutamyl peptides were chemically synthesized and purified using well-known techniques. Human CaSR cRNA was prepared from human kidney cDNA (Clontech) by a PCR method, as reported previously (Ohsu et al. 2010).

8.1.2.2 Determination of CaSR Activity Using HEK-293 Cells

Human CaSR cDNA was inserted into the pcDNA3.1 expression vector for transient transfection of HEK-293 cells. Briefly, CaSR cDNA was diluted in Opti-MEM I medium (Invitrogen), mixed with FuGENE 6 (Roche Applied Science), and added to subconfluent cultured HEK-293 cells. After incubation in a 96-well plate for 24 h, the cells were incubated with 5 μm Calcium-4 (Calcium-4 assay kit, Molecular Devices) for 45–60 min, and measurements were performed by using an image analyzer (FlexStation, Molecular Devices) and its associated software. Activation of CaSR expressed in HEK-293 cells leads to an increase in intercellular calcium ions. This was detected by using a calcium dye (Calcium-4) that binds the free Ca^{2+} and shows an increase in fluorescence, with excitation at 485 nm and emission at 525 nm. Concentration dependence of the fluorescence intensity was analyzed by using various CaSR agonists with or without 0.0002% (4.5 μm) NPS-2143. The assay buffer for calcium imaging contained 0.9 mm $CaCl_2$.

8.1.2.3 Human Sensory Analysis of CaSR Agonists

Samples were dissolved in distilled water and adjusted to pH 6.8–7.2 with NaOH. Human sensory analysis of a sample solution was performed by a previously described method (Ueda et al. 1990). A well-trained panel of 20 assessors was asked to evaluate the sample solutions using a 5-point rating scale, from −2 (apparently suppressed) to +2 (apparently strong). The known CaSR agonists were dissolved in a solution containing 0.1% MSG and 0.5% NaCl, and the taste intensity 5 s after tasting was evaluated and scored. Results were analyzed by paired t-test. The known CaSR agonists and an antagonist were used at the following final concentrations: calcium lactate (0.35%), protamine (0.02%), polylysine (0.08%), and L-histidine (0.2%). Cinacalcet was dissolved at a concentration of 1% in 99.5% ethanol and then diluted to 0.0015% (38 μM) with the tasting solution for sensory analysis. NPS-2143 was dissolved at a concentration of 0.1% in 99.5% ethanol and then diluted to 0.0002% (4.5 μM) with the tasting solution. The remaining ethanol in these tasting solutions did not affect sensory evaluation. Methods used in the human sensory analyses were approved by the Management Committee of the Food Product Application Center at Ajinomoto Co., Inc., and informed consent was obtained from all assessors.

8.1.2.4 Quantitative Analysis of the *Kokumi* Intensity Compared with GSH

The sensory activity as *kokumi* substances was evaluated as a point of subjective equivalence (PSE) as follows. The sensory analyses were performed by a panel of 17 well-trained assessors. Test samples of γ-Glu-Cys (0.15%), γ-Glu-Val (0.15%), γ-Glu-Ala (0.5%), γ-Glu-Abu-Gly (0.05%), or γ-Glu-Val-Gly (0.01%) were mixed with umami and salty taste solutions (0.05% MSG, 0.05% IMP, and 0.5% NaCl), and their tastes were compared with those of various concentrations of GSH as reference solutions. Eight concentrations of GSH in logarithmically equal steps at 50% intervals were used (0.02, 0.03, 0.044, 0.07, 0.10, 0.15, 0.23, and 0.34%, w/v). Each test sample was paired twice with each reference solution. Assessors judged the PSE between the sample and reference solutions. All judgments were dichotomous, i.e., the assessors were required to rate a test sample as either more or less intense than the reference GSH solution. The PSE was defined as the concentration that was judged to give a sensation that was equivalent to that of the standard solution. The analyses were performed by a panel of 17 well-trained assessors.

8.1.3 Results

8.1.3.1 Glutathione, a *Kokumi* Substance, Activates the Human CaSR

The effect of glutathione (GSH; γ-Glu-Cys-Gly) on human CaSR was investigated using CaSR-expressing HEK-293 cells. As shown in Fig. 8.1, glutathione activated human CaSR, and the EC_{50} value was calculated as 0.71 μM. In this study, the EC_{50} value of calcium ion was measured as 1.17 mM. From these results, glutathione is a potent CaSR agonist.

8.1.3.2 All Known CaSR Agonists Act as *Kokumi* Substance

The sensory properties of known CaSR agonists were investigated by human sensory analysis. Although calcium lactate, protamine, polylysine and L-histidine are used as food additives, it has been unclear whether these substances have an intrinsic taste. Cinacalcet, a synthetic allosteric CaSR activator, is used for the treatment of secondary hyperparathyroidism and hypercalcemia. Because some of the test substances had a bitter taste at higher concentrations, a suitable concentration without bitterness was selected for this study. As shown in Fig. 8.2, calcium lactate, protamine, polylysine, L-histidine, and cinacalcet showed significant taste enhancement at concentrations of 0.35, 0.02, 0.08, 0.2, and 0.0015% (w/v), respectively. In the sensory evaluation, the bitterness or other off-flavor was not detected when tested under our conditions.

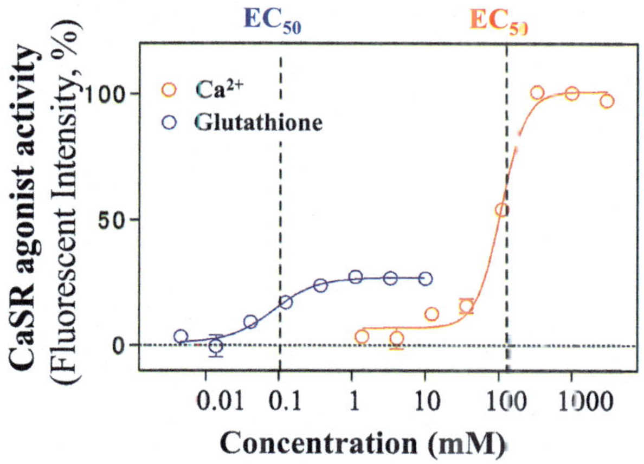

Fig. 8.1 CaSR activity of glutathione (GSH) and calcium ion measured using CaSR-expressing HEK-293 cells

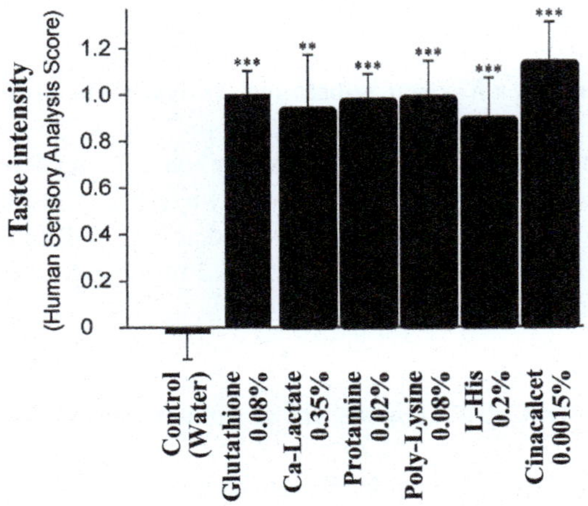

Fig. 8.2 Sensory activity as *kokumi* substance of known CaSR agonists. Human sensory analyses of the CaSR agonists were performed to determine the *kokumi* taste with known CaSR agonists. CaSR agonists were mixed with a control solution of MSG (0.1%) and NaCl (0.5%) at the following concentrations: GSH, 0.08% (control *kokumi* taste substance; calcium lactate, 0.35%; protamine, 0.02%; polylysine, 0.08%; L-histidine, 0.2%; cinacalcet, 0.0015%; γ-Glu-Val-Gly, 0.01%; and γ-Glu-Val-Leu, 0.08% (a low-activity control peptide). Human sensory analyses were performed in a blind test focusing on one of the *kokumi* taste characteristics, i.e., thickness. Data are shown as average scores ($n = 20$) with *error bars* (S.E.). *Asterisks* indicate significant differences (Student's *t*-test) from the control: ***, $p < 0.001$

8.1.3.3 CaSR Activity Is Positively Correlated with Sensory Activity as a *Kokumi* Substance

Various γ-glutamyl peptides were synthesized and were assessed for both CaSR activity and sensory activity as *kokumi* substances. Prior to evaluation, a total of 46 kinds of γ-glutamyl peptides were synthesized, and the CaSR activity was determined using *Xenopus* oocytes (Ohsu et al. 2010). Among the γ-glutamyl peptides tested, γ-Glu-Ala, γ-Glu-Val, γ-Glu-Cys, γ-Glu-Cys-Gly (GSH), γ-Glu-Abu-Gly (ophthalmic acid), and γ-Glu-Val-Gly were selected, and the CaSR activity was measured using CaSR-expressing HEK-293 cells, and EC_{50} values were calculated. The sensory activity as *kokumi* substances of these peptides were evaluated as a point of subjective equivalence (PSE) using glutathione (GSH) as the standard. The CaSR EC_{50} values of γ-Glu-Ala, γ-Glu-Val, γ-Glu-Cys, γ-Glu-Cys-Gly (GSH), γ-Glu-Abu-Gly (ophthalmic acid), and γ-Glu-Val-Gly were 3.7 μM, 1.6 μM, 0.46 μM, 0.71 μM, 0.025 μM, and 0.039 μM, respectively. The quantitative sensory evaluation test indicated that the glutathione concentrations for PSE of 0.5% γ-Glu-Ala, 0.15% γ-Glu-Val, 0.15% γ-Glu-Cys, 0.05% γ-Glu-Abu-Gly (ophthalmic acid), and 0.01% γ-Glu-Val-Gly were 0.074%, 0.092%, 0.094%, 0.085%, and 0.128%, respectively. The normalized sensory intensities (the ratio to glutathione intensity) as *kokumi* substance of γ-Glu-Ala, γ-Glu-Val, γ-Glu-Cys, γ-Glu-Cys-Gly (GSH), γ-Glu-Abu-Gly (ophthalmic acid), and γ-Glu-Val-Gly were 0.15, 0.61, 0.63, 1.7, and 12.8, respectively. Figure 8.3 displays the relation

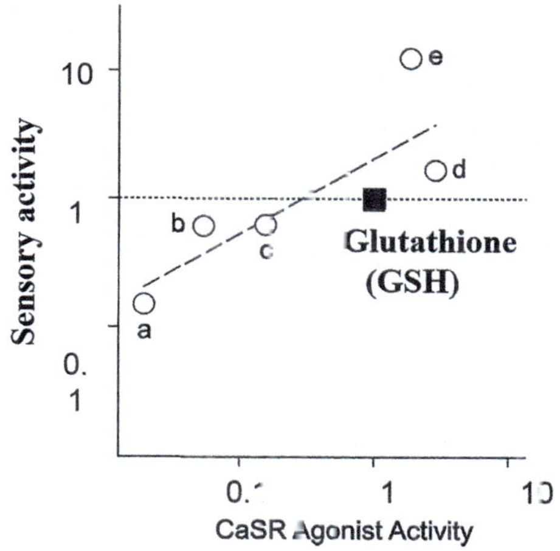

Fig. 8.3 The relationship between CaSR activity (*A*, EC_{50}) and *kokumi* taste intensity of five γ-glutamyl peptides is plotted on logarithmic scales in comparison with GSH. A Pearson correlation coefficient was used to calculate the relationship between CaSR activity and *kokumi* taste intensity. A significant positive correlation was found ($r^2 = 0.660$, $p = 0.0496$)

between the normalized CaSR activity and the normalized *kokumi* intensity, revealing that the CaSR activity of these γ-glutamyl peptides was positively correlated with the *kokumi* intensity ($r = 0.81$, $P < 0.05$).

8.1.3.4 A CaSR Antagonist Inhibits the Sensory Activity of Glutathione and γ-Glu-Val-Gly

In order to investigate the role of CaSR in the perception of *kokumi* substances, the influence of a CaSR antagonist, NPS-2143, on the sensory effect of two potent *kokumi* substances, GSH and γ-Glu-Val-Gly, was investigated. Figure 8.4 shows the effect of NPS-2143 on CaSR activation by GSH. Because NPS-2143 increased the EC_{50} value, it was shown to inhibit activation of CaSR by GSH. Human sensory analysis was performed to determine whether the *kokumi* intensity of GSH or γ-Glu-Val-Gly was suppressed in the presence of 0.0002% (4.5 μM) NPS-2143. As shown in Fig. 8.5a, the effectiveness of 0.08% GSH as a *kokumi* substance (*kokumi* intensity) for enhancing the taste of salt-umami solution was significantly reduced by addition of 0.0002% NPS-2143 ($p < 0.05$). As shown in Fig. 8.5b, *kokumi* intensity of 0.01% γ-Glu-Val-Gly was also significantly reduced by addition of 0.0002% NPS-2143 ($p < 0.001$). These results indicated that NPS-2143, a CaSR antagonist, effectively inhibited the sensory activity of both GSH and γ-Glu-Val-Gly.

8.1.4 Discussion

Glutathione (GSH) is a representative *kokumi* substance. Because GSH activates human CaSR (Fig. 8.1), we hypothesized that CaSR may be involved in the perception of *kokumi* substances by humans. To investigate this hypothesis, the

Fig. 8.4 Effect of CaSR inhibitor NPS-2143 on CaSR activity measured using CaSR-expressing HEK-293 cells

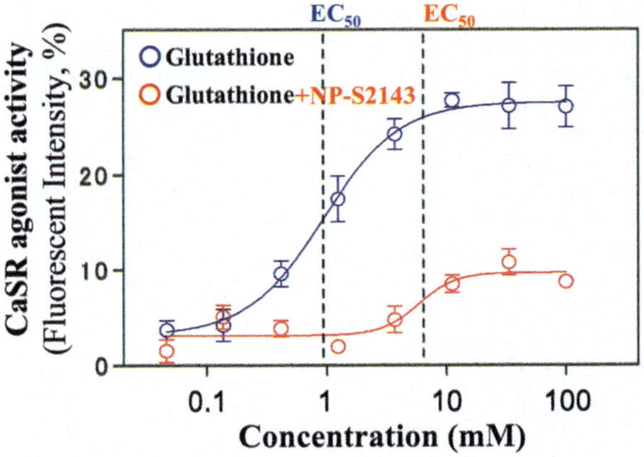

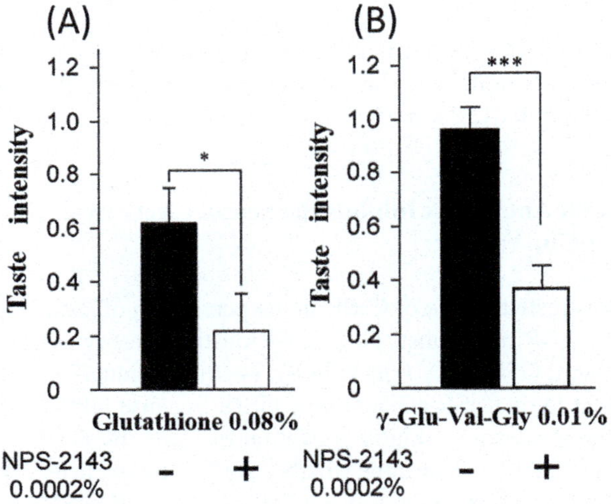

Fig. 8.5 Effect of CaSR inhibitor NPS-2143 on sensory effect of glutathione and γ-Glu-Val-Gly. Human sensory analyses of CaSR agonists were performed to determine *kokumi* taste in the presence or absence of a CaSR antagonist (NPS-2143) using the method described for Fig. 8.3a. The concentrations used in this experiment were as follows: GSH, 0.08%; γ-Glu-Val-Gly, 0.01%; and NPS-2143, 0.0002% (4.5 μm). Data are shown as average scores ($n = 20$) with *error bars* (S.E.). *Asterisks* indicate significant differences (Student's *t*-test) from NSP-2143: *, $p < 0.05$; ***, $p < 0.001$

CaSR-activating effect and sensory properties of various compounds were investigated. CaSR activation was assessed by using CaSR-expressing HEK-293 cells, while sensory properties were determined by human sensory analysis. In the sensory analysis, the test compounds were added to a salt-umami solution and taste enhancement at 5 s after tasting was evaluated. First, we investigated the activity of known CaSR agonists as *kokumi* substances. As shown in Fig. 8.2, all of the CaSR agonists that we tested had activity as *kokumi* substances, suggesting that CaSR is related to perception of *kokumi* substances by humans. Second, we used several γ-glutamyl peptides, such as γ-Glu-Ala, γ-Glu-Val, γ-Glu-Cys, GSH, γ-Glu-Abu-Gly, and γ-Glu-Val-Gly, to investigate the correlation between CaSR activity and sensory activity as a *kokumi* substance. Figure 8.3 shows that the extent of CaSR activation by these γ-glutamyl peptides had a significant positive correlation with their *kokumi* intensity ($r = 0.81$, $P < 0.05$). These results provided strong evidence for involvement of that CaSR in perception of *kokumi* substances. Third, the effect of NPS-2143, a CaSR-specific antagonist, on sensory activity as *kokumi* substances was investigated. As shown in Fig. 8.5a, the *kokumi* intensity (enhancement of the taste of a salt-umami solution) of 0.08% GSH was significantly reduced by addition of 0.0002% NPS-2143 ($p < 0.05$). Likewise, *kokumi* intensity of 0.01% γ-Glu-Val-Gly was significantly reduced by addition of 0.0002% NPS-2143 ($p < 0.001$, Fig. 8.5b). These results strongly suggest that CaSR is involved in perception of *kokumi* substances by humans.

Thus, we performed the following investigations: [1] the sensory evaluation of the property as *kokumi* substances of known CaSR agonists, [2] assessment of the correlation between CaSR activation and *kokumi* intensity of several γ-glutamyl peptides, and [3] determination of the effect of a CaSR antagonist on the *kokumi* intensity of GSH and γ-Glu-Val-Gly. All of our results clearly indicated that CaSR is involved in perception of *kokumi* substances by humans.

In the next study, to clarify the mechanisms involved in perception of *kokumi* substance, including the function of CaSR in taste cells, CaSR expression in taste cells was investigated, and the response of taste cells to *kokumi* substances was determined by using semi-intact mouse lingual slices. The possibility of coexpression of CaSR and T1R3 was also investigated.

8.2 Taste Cells Expressing Calcium-Sensing Receptor (CaSR) Respond to *Kokumi* Substances

Abstract Some foods are known to have flavor sensations, such as continuity, mouthfulness, and thick flavor, which cannot be explained by the five basic tastes alone. It has been demonstrated that these sensations are evoked by the addition of *kokumi* substances, which are flavor modifiers with no actual taste themselves. However, the mechanism is poorly understood. From the results obtained in the last section, it was suggested that CaSR is involved in perception of *kokumi* substances by humans. In this section, we investigated the receptor cells for *kokumi* substances and their physiological properties using Calcium Green-1-loaded mouse taste cells in lingual tissue slices and confocal microscopy. Focal application of *kokumi* substances around the taste pores induced an increase in the intracellular Ca^{2+} concentration ($[Ca^{2+}](i)$) in a subset of taste cells. This response was inhibited by pretreatment with a CaSR antagonist, NPS2143. Finally, CaSR-expressing taste cells were identified as a different cellular subset from the T1R3-expressing umami or sweet taste receptor cells. The present observations indicate that CaSR-expressing taste cells are the primary detectors of *kokumi* substances and exist as an independent population from basic taste receptor cells, at least for sweet and umami tastes.

8.2.1 Introduction

In the previous study, we showed that various CaSR agonists including γ-glutamyl-cysteinyl-glycine (the reduced form of glutathione, GSH) and other γ-glutamyl peptides have enhanced umami, sweet, and salty tastes and that there was a strong correlation between CaSR agonist activity and *kokumi* intensity. Ueda et al. reported that aqueous extracts of garlic containing GSH enhanced umami taste intensity, and they proposed that the taste-enhancing substances should be called "*kokumi* substances" (Ueda et al. 1990; Ueda et al. 1997). We previously identified several γ-glutamyl peptides that were CaSR agonists and had a property as *kokumi* substances, among which γ-glutamyl-valinyl-glycine (γ-Glu-Val-Gly) was the most potent *kokumi* substance (Ohsu et al. 2010). These findings suggested that CaSR-expressing taste cells in the lingual epithelium may respond to *kokumi* substances.

In the present study, we employed a semi-intact lingual slice preparation for focal application of *kokumi* substances to the apical tips of the taste buds, and we measured individual cellular responses with sufficient time and spatial resolution for Ca^{2+} imaging.

8.2.2 Materials and Methods

8.2.2.1 Tissue Preparation and Functional Imaging

All experimental procedures were approved by the Animal Experiment Institution Review Board of Ajinomoto Co., Inc., Institute for Innovation (2,008,220, 2,009,085, 2,010,013, and 2,011,239) and conformed with the standards for the use of laboratory animals published by the Institute of Laboratory Animal Resources, U.S. National Academy of Sciences. Male C57BL/6 adult mice (≥7 weeks old) were sacrificed by exposure to diethyl ether, followed by cervical dislocation. The tongues were removed and immersed in cold Tyrode's solution. Lingual slices containing the vallate papilla were obtained, and a Ca^{2+} indicator dye was injected into the taste cells by the procedure previously described by Maruyama et al. (2006). Briefly, a fluorescent Ca^{2+} indicator dye, Calcium Green-1 dextran (CGD; MW 3000; 0.25 mM in H_2O; Invitrogen, Carlsbad, CA, USA), was injected iontophoretically through a large-diameter-tip glass micropipette (40 μm) into the crypt surrounding the vallate papilla (−3.5 μA square pulses, 10 min). The CGD-loaded tissue was sliced at 100 μm with a vibratome (Leica VT1000S, Nussloch, Germany). Slices containing vallate taste buds were mounted on a glass coverslip coated with Cell-Tak (Becton Dickinson, Franklin Lakes, NJ, USA), put in a recording chamber, and superfused with Tyrode's solution (30 °C) at a rate of 1.5 ml/min. Single glass micropipettes (2-μm tip diameter) were used to directly deliver taste stimuli for apical stimulation of a selected taste bud. Stimuli were ejected for 1 s with air pressure (3.5 psi; Pressure System IIe, Toohey Company, Fairfield, NJ, USA). Different pipettes were mounted for each taste stimulus. All stimulus solutions contained 2 μM fluorescein to monitor stimulus application, duration, and concentration.

CGD-loaded taste cells were viewed with a laser-scanning confocal microscope, using an argon laser (Fluoview FV-300, Olympus, Tokyo, Japan). Images were captured at 1.1 s intervals. Fluorometric signals are expressed as relative fluorescence change: $\Delta F/F = (F-F_0)/F_0$, where F_0 denotes the resting fluorescence level corrected for any bleaching that occurred during the recording. Using $\Delta F/F$ corrects for variations in baseline fluorescence, cell thickness, total dye concentration, and illumination (Helmchen 2000). Peak $\Delta F/F$ constituted the response amplitude for statistical quantification.

8.2.2.2 Data Analysis

Statistical analyses using paired Student's *t*-tests were applied to determine whether the changes in the response amplitudes (peak $\Delta F/F$) to a given treatment were significant. Data presented in bar graphs show mean ± SEM.

8.2.2.3 Reagents and Solutions

γ-Glutamyl-valinyl-glycine (γ-Glu-Val-Gly) was synthesized by Kokusan Chemicals (Tokyo, Japan). Cinacalcet, NPS2143, and SC-45647 were chemically synthesized in our facility by previously described methods (Rodriguez et al. 2005; Rybczynska et al. 2006; Nofre et al. 1987). All other chemicals including monopotassium L-glutamate (MPG) were purchased from Sigma Chemical (St Louis, MO, USA). All tastants were freshly dissolved in Tyrode's solution for each experiment. The standard medium consisted of Tyrode's solution, which is composed of, in mM: 135 NaCl, 5 KCl, 1.5 $CaCl_2$, 1 $MgCl_2$, 10 HEPES, 10 glucose, 10 sodium pyruvate, and 5 $NaHCO_3$, pH 7.2; 318–323 mOsm. For Ca^{2+}-free Tyrode's solution, $CaCl_2$ was removed, and 0.2 mM EGTA was added.

8.2.2.4 Immunohistochemistry

Mouse circumvallate tissues were fixed in 4% paraformaldehyde and cryoprotected in 10–30% sucrose for 3.5 h at 4 °C. Frozen sections (12 μm) were prepared and blocked in Protein Block (Dako, Glostrup, Denmark) with 1% Triton X-100 (Sigma) for 45 min at room temperature. Sections were then incubated for 1 h with a primary antibody [rat anti-CaSR, diluted 1:400; rabbit anti-PLCβ2, 1:500 (sc-206, Santa Cruz, Santa Cruz, CA, USA); rabbit anti-NCAM, 1:400 (AB5032, Millipore, Billerica, MA, USA); and rabbit anti-T1R3, 1:800 (kind gift from Dr. Iwatsuki, Ajinomoto Co., Inc.) (Iwatsuki et al. 2009)], followed by a secondary antibody (1:1000): Alexa Fluor 488 labeled donkey anti-rat IgG antibody for anti-CaSR (A-21208, Invitrogen) and Alexa Fluor 568 labeled goat anti-rabbit IgG antibody for anti-PLCβ2, anti-NCAM, and anti-T1R3 (A-11036, Invitrogen). CaSR was detected using a specific antibody generated in our lab (host = rat). The anti-mouse-CaSR antibody recognizes the protein sequences KSNSEDPFPQPERQKQQ, QGPMVGDHQPEIESPDE, and MRQNSLEAQKSNDTLNR, corresponding to residues 917–933, 1034–1050, and 994–1010 of the mouse CaSR, respectively. Negative controls without the primary antibody were processed in parallel in every experiment. Images were obtained with a laser-scanning confocal microscope (Olympus). We estimated a thickness of ~3 μm for the optical section taken by the confocal microscope.

8.2.2.5 RT-PCR Analysis

RT-PCR amplification was performed using primers that amplify the mouse *CaSR*, *PLCβ2*, *Snap25*, and *β-actin*. Briefly, dissected tongue containing circumvallate and foliate papillae was injected into the submucosal layer with a mixture of 1 mg/ml collagenase A (Roche Applied Science, Indianapolis, IN, USA), 2.5 mg/ml Dispase II (Roche Applied Science), and 1 mg/ml trypsin inhibitor (Sigma) and then incubated for 20 min at room temperature. The papilla-containing epithelium was peeled from the underlying connective tissue. Total RNA was isolated from the epithelial papillae and from the epithelium without taste buds (RNA Micro Kit, Agilent Technologies, Santa Clara, CA, USA). Purified RNA was denatured, and first-strand cDNA was synthesized using oligo$(dT)_{12-18}$ primer and reverse transcriptase (Super Script III, Invitrogen). cDNA was used as a template in a 20 μl PCR mixture with Taq polymerase (Invitrogen). PCR conditions were as follows: 94 °C for 2 min, followed by 29–35 cycles of 94 °C for 30 s, 58 °C for 20 s, and 72 °C for 45 s. The PCR products were analyzed by gel electrophoresis with GelRed staining (Biotium, Hayward, CA, USA). The primers used were as follows: *CaSR*, 5′-tcgagaccccttacatggac-3′ (forward) and 5′-agtagttccccaccaggtca-3′ (reverse); *PLCβ2*, 5′-ctcgctttgggaagtttgc-3′ (forward) and 5′-gcattgactgtcatcgggt-3′ (reverse); *Snap25*, 5′-ggcaataatcaggatggagtag-3′ (forward) and 5′-agatttaaccacttcccagca-3′ (reverse); and *β-actin*, 5′-caccctgtgctgctcacc-3′ (forward) and 5′-gcacgatttccctctcag-3′ (reverse).

8.2.3 Results

8.2.3.1 *CaSR* Is Expressed in the Taste Buds in Lingual Epithelia

We tested the expression of *CaSR* mRNA in taste buds and in non-taste lingual epithelium from a C57BL/6 mouse by RT-PCR. We confirmed that *CaSR* mRNA was expressed in circumvallate and foliate, but not in non-taste epithelium (Fig. 8.6a). To determine the presence of CaSR in taste cells, we employed immunohistochemistry on mice lingual tissues. CaSR immunoreactivity was observed in a subset of spindle-shaped taste cells in circumvallate, foliate, fungiform and palate papillae (Fig. 8.6b–e). In the transverse section of circumvallate taste buds, 8–10 CaSR-immunoreactive taste cells were present in a taste bud (Fig. 8.7d, h). The specificity of the antibody was confirmed by antigen preabsorption, which resulted in little or no immunoreaction in taste cells (Fig. 8.6f).

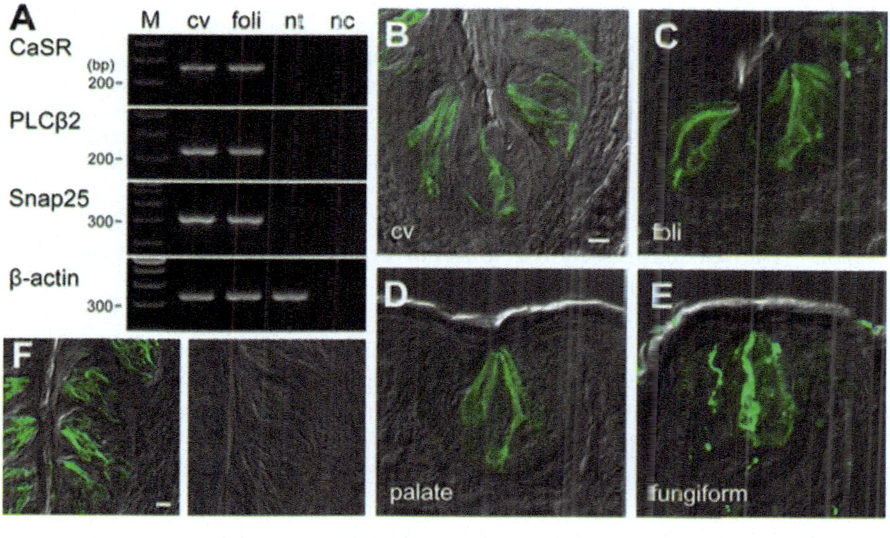

Fig. 8.6 Taste cells express CaSR. (**a**) RT-PCR for *CaSR* expression in taste bud-enriched circumvallate (cv), foliate (foli) and non-taste bud (nt) lingual epithelium nc negative control (lacking template); M, molecular standard. (**b**–**e**) Immunostaining for CaSR in taste buds. CaSR immunofluorescence is seen in most circumvallate (**b**), foliate (**c**), palate (**d**), and fungiform (**e**) taste buds. Immunofluorescent images (green) were superimposed on DIC images. (**f**) Validating the anti-CaSR antibody. The CaSR antiserum was preabsorbed with an excess of antigen peptides. The circumvallate sections reacted with preabsorbed and non-absorbed antibodies and were processed simultaneously. Images were taken under the same illumination conditions and detector settings. Scale bars 20 μm

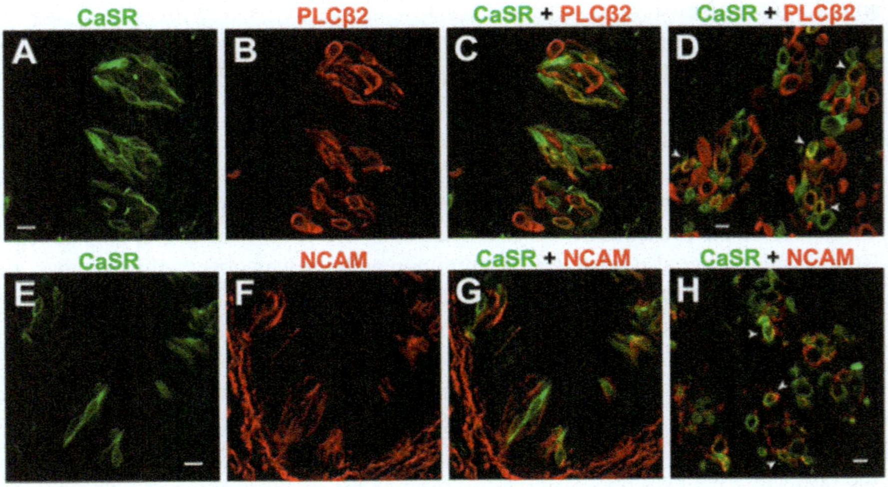

Fig. 8.7 Confocal images showing colocalization of CaSR and the taste cell markers in taste cells from mouse circumvallate papillae. (**a–c**) A longitudinal section of a circumvallate taste bud immunostained with antibodies against CaSR (**a**) and PLCβ2 (**b**). (**c**) Overlay of A and B. (**d**) A transverse section of a circumvallate taste bud immunostained with antibodies against CaSR (green) and PLCβ2 (red). (**e–g**) A longitudinal section of a circumvallate taste bud immunostained with antibodies against CaSR (**e**) and NCAM (**f**). (**g**) Overlay of (**e**) and (**f**). (**h**) A transverse section of a circumvallate taste bud immunostained with antibodies against CaSR (green) and NCAM (red). Scale bars 20 μm. Arrowheads indicate double-labeled cells

8.2.3.2 CaSR Is Expressed in a Subset of Type II (Receptor) and Type III (Presynaptic) Cells

Mammalian taste buds contain three distinct classes of cells (Chaudhari and Roper 2010, Murray 1993; Yee et al. 2001). A heterogeneous population of mammalian taste cells includes morphologically and functionally different taste cells classified into three subtypes, type I (glial-like cells), type II (receptor cells), and type III (presynaptic) taste cells (Kinnamon et al. 1985). These classes express different complements of genes related to their functions: receptor (Type II) cells express G-protein-coupled taste receptors and transduction machinery. In contrast, presynaptic (Type III) cells express neuronal proteins, including those associated with synapses, and also respond to sour stimuli (DeFazio et al. 2006; Huang et al. 2006; Huang et al. 2008; Medler et al. 2003; Tomchik et al. 2007).

To characterize the CaSR-immunoreactive taste cells, we investigated the coexpression of CaSR and taste cell markers. Using immunofluorescence, previous reports have shown in circumvallate taste buds the expression of PLCβ2 in type II taste cells (receptor cells) (DeFazio et al. 2006) and of neural cell adhesion molecule (NCAM) in type III taste cells (presynaptic cells) (Dvoryanchikov et al. 2007). Consequently, we used double immunofluorescence microscopy to evaluate whether these CaSR-expressing taste cells coexpressed taste cell markers. Using immunohistochemistry, we observed that the CaSR-expressing taste cells also expressed either PLCβ2 or NCAM (Fig. 8.7). Out of 728 CaSR-positive cells, 314 cells expressed PLCβ2 (43.1%; Fig. 8.7d), while the other CaSR-positive cells expressed NCAM (669 cells out of 1033 cells, 64.7%; Fig. 8.7h). Conversely, out of 823 PLCβ2-positive cells, 314 cells expressed CaSR (38.2%). These populations were in agreement with previous findings regarding the cell types of CaSR-expressing taste cells (San Gabriel et al. 2009).

8.2.3.3 Taste Cells Respond to Focally Applied *Kokumi* Substances

To test whether *kokumi* substances induce an intracellular Ca^{2+} response in taste cells and, if so, to identify which taste cells are responsible, we employed semi-intact lingual slice preparations, focally applied *kokumi* substances to the apical chemosensory tips of the taste cells, and imaged the intracellular [Ca^{2+}] changes in the taste cells with confocal scanning microscopy (Caicedo et al. 2000; Maruyama et al. 2006). Focal application of the *kokumi* substances, cinacalcet (a classic CaSR agonist; 10 μM), glutathione (GSH; 100 μM), or γ-Glu-Val-Gly (100 μM) induced Ca^{2+} responses ($\Delta[Ca^{2+}]_i$) in a small fraction of the taste cells (Fig. 8.8a) (Ohsu et al. 2010). These responses were not induced by solution puffing, because no response was observed after ejection of Tyrode's solution (Fig. 8.8a). Some, but not all, *kokumi* substance-responsive cells also responded to bath applied KCl (50 mM), which induces a Ca^{2+} response in the presynaptic (Type III) taste cells via depolarization of the plasma membrane. γ-Glu-Val-Gly evoked a transient $[Ca^{2+}]_i$ increase in 6.5% of the taste cells in the circumvallate papilla (34 of 524 cells obtained from 26 mice). The mean amplitude of Ca^{2+} responses ($\Delta F/F$) evoked by 100 μM γ-Glu-Val-Gly was 11.6 ± 1.7% (mean ± SE; n = 21 cells; Fig. 8.8b). The EC_{50} value of γ-Glu-Val-Gly was estimated at approximately 13 μM (Fig. 8.8b). Importantly, applying higher concentrations of γ-Glu-Val-Gly (>30 μM) did not induce Ca^{2+} responses in additional taste cells. Furthermore, *kokumi* substance-induced responses were selectively blocked by 3 μM NPS2143, a CaSR inhibitor, which barely affected the umami (MPG 100 mM + IMP 1 mM) and the sweet (SC45647, 100 μM) responses (Fig. 8.8c) (Ohsu et al. 2010; Gowen et al. 2000). This suggests that the Ca^{2+} responses we recorded reflect selective stimulation of a specific subpopulation of *kokumi* substance-responsive taste cells.

8.2.3.4 Responses to *Kokumi* Substance Stimulus Involve Ca^{2+} Release

Next, we investigated the Ca^{2+}-mobilizing pathway that is activated by the *kokumi* substance stimulus in mouse circumvallate taste cells. We examined responses in acute absence of extracellular Ca^{2+} by bathing slices in a Ca^{2+}-free Tyrode's solution (containing 0.2 mM EGTA) 2 min before the focal *kokumi* substance stimulation. As shown in Fig. 8.9a, responses evoked by γ-Glu-Val-Gly did not significantly change compared with the presence of extracellular Ca^{2+} (control, $\Delta F/F = 7.1 \pm 1.8\%$;

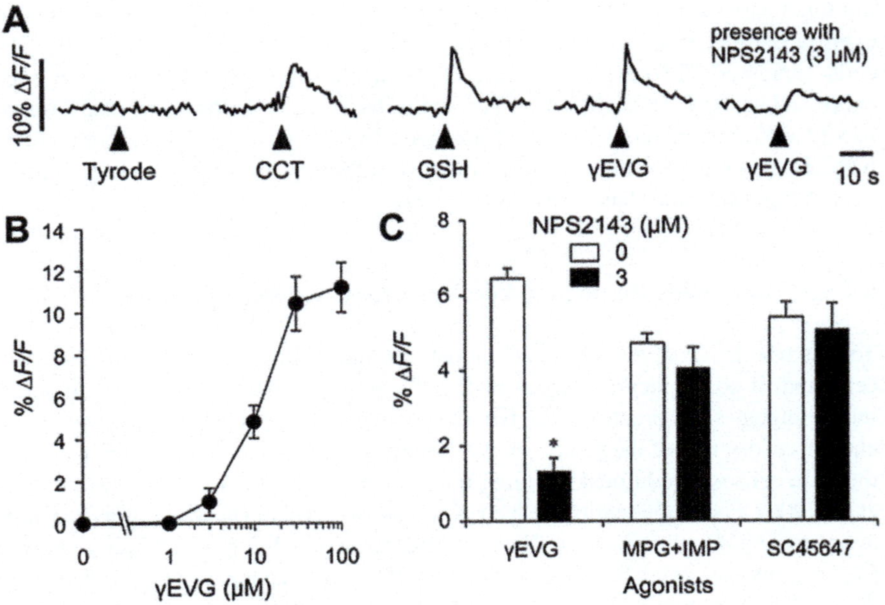

Fig. 8.8 Taste cell responses (ΔCa^{2+}) evoked by *kokumi* stimuli, recorded in a slice preparation of the mouse circumvallate papilla. (**a**) Taste cells were stimulated sequentially with three *kokumi* substances, cinacalcet (CCT, 10 μM), glutathione (GSH, 100 μM), and γ-glutamyl-valinyl-glycine (γEVG, 100 μM), as well as γEVG + NPS2143 (3 μM), a CaSR antagonist. Arrowheads below traces indicate the stimulation. (**b**) Concentration-response relationship for γEVG (mean ± SE; n = 4 cells). (**c**) Taste responses elicited by γEVG (100 μM) were inhibited by the CaSR antagonist, NPS2143 (3 μM), but the umami (MPG 100 mM + IMP 1 mM) and the sweet (SC45647, 10 μM) responses were unaffected. Mean amplitudes of γEVG-, MPG + IMP-, and SC45647-induced responses in the presence or absence of 3 μM NPS2143 are shown (mean ± SE; $*p \leq 0.05$, n = 4 cells). Raw traces are shown in A

Ca^{2+}-free, $\Delta F/F = 6.7 \pm 2.0\%$; n = 5). In contrast, the depolarization-evoked Ca^{2+} response, elicited by perfusing the slice with 50 mM KCl that allows Ca^{2+} influx through voltage-gated Ca^{2+} channels in type III presynaptic taste cells (Richter et al. 2004), was nearly completely abolished in the absence of extracellular Ca^{2+} under parallel treatments (control, $\Delta F/F = 16.1 \pm 3.6\%$; Ca^{2+}-free, $\Delta F/F = 2.8 \pm 0.2\%$; $n = 4$; Fig. 8.9b). Part of the γ-Glu-Val-Gly-responsive taste cells showed a Ca^{2+} response to KCl stimulation; however, the γ-Glu-Val-Gly-induced response in these cells was not affected by Ca^{2+}-free conditions (Fig. 8.9a, b). These results are consistent with *kokumi* transduction involving release of stored Ca^{2+}, not Ca^{2+} influx (Fig. 8.9c).

Umami, sweet, and bitter stimuli trigger the release of stored Ca^{2+} by activating phospholipase C (PLC) (Maruyama et al. 2006; Huang et al. 1999; Ogura and Kinnamon 1999; Rossler et al. 1998; Zhang et al. 2003). To directly test whether *kokumi* substance-elicited Ca^{2+} responses arise from PLC activation, we used a non-selective PLC inhibitor, U73122 (Bleasdale et al. 1990; Salari et al. 1993; Thompson

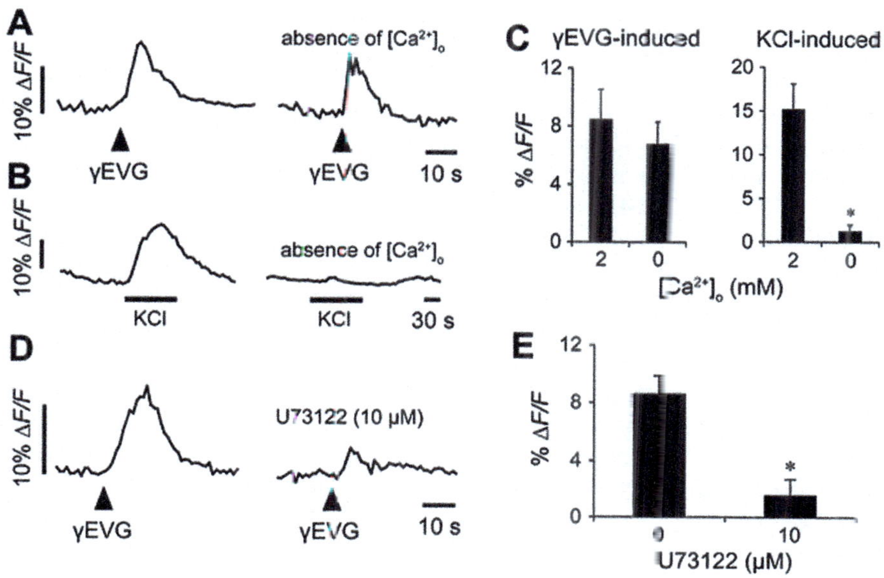

Fig. 8.9 Ca^{2+} response elicited by γEVG involves intracellular Ca^{2+} stores and phospholipase C. (**a**) γEVG (100 μM) was focally applied in medium containing Ca^{2+} (left trace) or in the absence of Ca^{2+} (Ca-free medium with 0.2 mM EGTA; right trace). (**b**) Responses elicited by depolarization (bath-applied KCl, 50 mM) and influx of Ca^{2+} through voltage-dependent Ca^{2+} channels were abolished in the absence of extracellular Ca^{2+}. (**c**) Mean amplitudes of responses in the presence or absence of Ca^{2+} in the medium (mean ± SE; $*p \leq 0.05$, n = 4 cells). (**d**) Responses to γEVG inhibited by U73122 (10 μM). (**e**) Mean amplitudes of the responses in the presence or absence of U73122 (mean ± SE; $*p \leq 0.05$, n = 4 cells)

et al. 1991). After incubation with 10 μM U73122 for 10 min, responses elicited by γEVG were almost abolished (control, $\Delta F/F$ = 8.3 ± 1.6%; U73122, $\Delta F/F$ = 1.7 ± 0.8%; n = 4; Fig. 8.9d, e). In contrast, depolarization (KCl)-induced responses were not significantly altered by treatment with U73122 (data not shown). These data strongly support the notion that the *kokumi* mechanisms involve intracellular Ca^{2+} release.

8.2.3.5 CaSR Ligand-Responsive Cells Do Not Respond to L-Glutamate Stimuli

It has been reported that CaSR is activated by various γ-glutamyl peptides, including glutathione and γ-Glu-Val-Gly (Ohsu et al. 2010; Conigrave et al. 2000; Wang et al. 2006). When transiently expressed in HEK293 cells, CaSR also induces Ca^{2+} changes in response to L-glutamate monomer, which is associated with the umami taste (Bystrova et al. 2010). In contrast, receptor (type II) cells respond to sweet, bitter, and umami taste stimuli by elevating cytoplasmic Ca^{2+} (DeFazio et al. 2006;

Tomchik et al. 2007). We asked whether the CaSR ligand, γ-Glu-Val-Gly, and glutamate produce responses in the same mouse type II taste cells. These studies have suggested that γ-Glu-Val-Gly mimics the taste of L-glutamate, at least in part, by activating the same taste receptors as umami compounds. To test this interpretation directly, we focally applied γ-Glu-Val-Gly and monopotassium L-glutamate (MPG) + inosine monophosphate (IMP) sequentially on circumvallate taste buds. γ-Glu-Val-Gly (100 μM) evoked transient Ca^{2+} responses in some taste cells (10 responding cells out of 132 recorded cells; $\Delta F/F = 7.4 \pm 1.3\%$), but not in those that responded to MPG (100 mM) + IMP (1 mM) (Fig. 8.10a, c). Conversely, MPG + IMP-responding cells (8 cells out of 132 recorded cells; $\Delta F/F = 6.7 \pm 1.5\%$) did not respond to γ-Glu-Val-Gly (Fig. 8.10b, d). These data suggest that separate receptors, found on separate cells, generate Ca^{2+} responses to γ-Glu-Val-Gly and MPG + IMP (Fig. 8.10c). In cells that respond to each agonist, we cannot rule out the possibility of subthreshold responses to the other agonist. Nevertheless, these results emphasize that responses to CaSR ligand in native taste tissues are highly heterogeneous and vary markedly from those described for the proposed umami taste receptors.

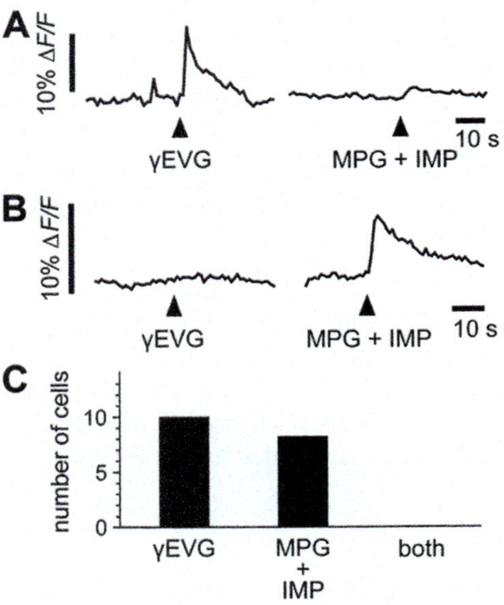

Fig. 8.10 γ-Glu-Val-Gly-responding taste cells are different from MPG + IMP-responding cells. (**a**) Taste cell responses were recorded in lingual slice preparations of mouse circumvallate papilla. Preparations were sequentially stimulated with γ-Glu-Val-Gly (100 μM) and MPG (100 mM) + IMP (1 mM). The traces show superimposed responses from γ-Glu-Val-Gly-responding cells (**a**) and MPG + IMP-responding cells (**b**). Responses to γ-Glu-Val-Gly were only observed in cells lacking response to MPG + IMP. (**c**) We recorded from 132 Calcium Green-loaded taste cells in 16 lingual slices. Ten dye-loaded cells were γ-Glu-Val-Gly-responding cells, while MPG + IMP stimulation evoked responses in 8 out of 132 cells. We did not identify cells that responded to both substances

8.2.3.6 Physiological Responses Correlate with Molecular Expression

We recorded Ca^{2+} responses to 100 μM γ-Glu-Val-Gly and 100 nM MPG in lingual slice preparations as described above. The functional responses of taste cells in lingual slices fell into two distinct classes. Our next step was to test whether the two kinds of responding taste cells, determined by functional imaging, mapped onto the two categories determined by expression of CaSR and an umami receptor subunit, T1R3 (Li et al. 2002). We designed additional methods, which are independent from the methods described above, to distinguish between CaSR-expressing cells and T1R3 cells. To identify these receptors, we used dual immunohistochemistry for CaSR and T1R3 in mouse circumvallate papillae. Examples of immunostained circumvallate papilla are shown in Fig. 8.11. The presence of CaSR immunofluorescent signals was observed in a subset of taste cells. In 63 circumvallate taste buds, 502 taste cells expressed CaSR, whereas 347 taste cells expressed T1R3. Only three cells (0.6% of CaSR-positive taste cells) expressed both CaSR and T1R3 (Fig. 8.11). These data demonstrate that most taste cells express CaSR, T1R3, or neither.

8.2.4 Discussion

In this study, we used a preparation of lingual slices from circumvallate papilla that allows one to apply tastants selectively to the apical chemosensory tips of taste cells while avoiding stimulating non-taste cells and basolateral regions of taste buds (Maruyama et al. 2006). This report presents data that indicate that CaSR ligands

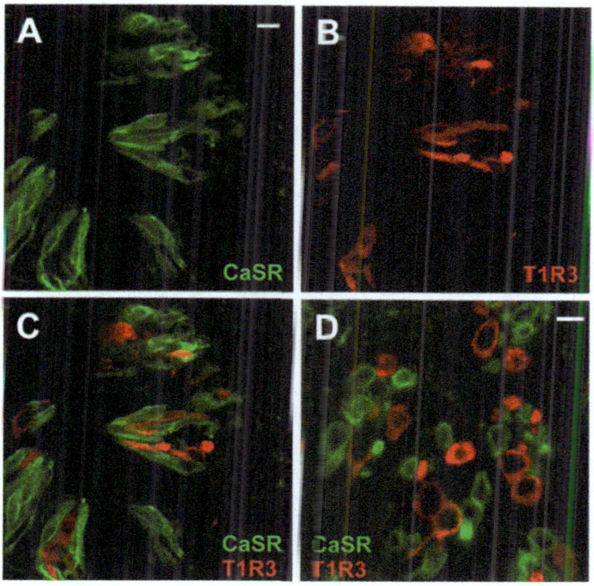

Fig. 8.11 CaSR is found in distinct cells that do not express an umami/sweet receptor subunit. (**a–c**) A longitudinal section of a circumvallate taste bud immunostained with antibodies against CaSR (**a**) and T1R3 (**b**). (**c**) Overlay of A and B. (**d**) A transverse section of a circumvallate taste bud immunostained with antibodies against CaSR (green) and T1R3 (red). Scale bars 20 μm

induce a cellular response in mouse circumvallate taste cells. We identified a population (6.5%) of taste cells in the circumvallate papillae that responds to *kokumi* substances. This value is comparable with the immunohistochemistry results (6.8% CaSR-positive cells/taste bud). Furthermore, this compares with previous estimates, in the same preparation, of 28% cells that respond to bitter (Caicedo et al. 2002) and 5% to umami (monopotassium L-glutamate) stimuli (Maruyama et al. 2006). The threshold concentration of γ-Glu-Val-Gly needed to activate a Ca^{2+} response in taste cells was 3 μM, which is consistent with reports on human sensory evaluation analysis (Ohsu et al. 2010). We also observed that both species respond to CaSR agonists with very similar characteristics (Maruyama et al. 2012). Hence, the results from mouse experiments can be extrapolated to results from human CaSR experiments. Additionally, NPS2143, a CaSR inhibitor, impaired the CaSR agonist-induced taste cell response in lingual slice preparations, as well as in heterologous experiments. Moreover, taste cells responded to multiple CaSR agonists, including hydrophobic (cinacalcet) and hydrophilic (GSH and γ-Glu-Val-Gly) *kokumi* substances. Taste buds contain tight junctions around the taste pore (e.g., claudin 4 and 8) and maintain a highly specific permeability barrier for paracellular diffusion (Michlig et al. 2007). It is expected that CaSR agonists (*kokumi* substances), especially the hydrophilic GSH and γ-Glu-Val-Gly, are accessible only to microvilli in the apical tip of the taste cells and activate CaSR. Taken together, the data from the present study suggest that CaSR is involved in taste transduction in mice.

Human sensory analysis demonstrated that CaSR agonists enhance the intensity of umami and sweet tastes and are called *kokumi* flavor (Ohsu et al. 2010). We expected that CaSR agonists would induce a response in the umami or sweet taste cells. Tordoff and colleagues reported that calcium elicits appetitive behavior in mice (McCaughey et al. 2005). Their recent study suggested that CaSR dimerizes with the sweet and umami receptor subunit T1R3 (Tordoff et al. 2008). However, surprisingly, CaSR agonists induced a response in a different group of taste cells from the umami or sweet taste cells. Results from double immunohistochemistry of CaSR and T1R3 also support this observation. Signaling in taste cells, especially for umami, sweet, and bitter, are well understood. Taste receptors for umami (T1R1 + T1R3) and sweet (T1R2 + T1R3) agonists have been identified (Li et al. 2002; Montmayeur et al. 2001; Nelson et al. 2001; Nelson et al. 2002), and activation of these receptors elicits cellular responses (e.g., transient Ca^{2+} changes and ATP secretion) in receptor-expressing taste cells. These findings demonstrate the involvement of CaSR in *kokumi* signaling and suggest that CaSR is not directly involved in umami or sweet taste signaling. In cells that respond to each agonist, we cannot rule out the possibility of cell-to-cell signaling within a taste bud.

Our results indicate that *kokumi* substance-responsive taste cells are both presynaptic (type III) and non-presynaptic taste cells. Furthermore, the results suggest that certain stimuli evoke the Ca^{2+} response via release from intracellular stores in both presynaptic and non-presynaptic taste cells. The exact classification of *kokumi* substance-responsive taste cells remains to be determined. San Gabriel et al. reported that CaSR-positive taste cells coexpress PLCβ2, a receptor (type II) cell marker, or NCAM, a presynaptic cell marker in rat vallate taste buds San Gabriel et al. 2009). Taken together with these observations, CaSR is expressed at least in type II and type III taste cells. Activation of *kokumi* substance-responsive cells may modulate the activity of sensory afferent fibers and/or neighboring taste cells in the same taste bud (Maruyama et al. 2006; Huang et al. 2005; Huang et al. 2007). ATP released from receptor cells may excite primary sensory afferent fibers (Finger et al. 2005). ATP may also function as a paracrine transmitter and act on cells within the taste bud (Huang et al. 2007; Dando and Roper 2009; Romanov et al. 2007). Our experiments were not designed to distinguish between these possibilities, both of which remain open questions.

Recently, Bystrova et al. reported that only type III taste cells responded to agonists of CaSR. However, in our study, all CaSR agonists tested induced responses in both type II and type III taste cells. We do not know the reason for this discrepancy; enzyme treatment during taste cell isolation might affect cellular responses. In PLCβ2-positive type II taste cells, CaSR is expressed in T1R3-negative taste cells. Whether CaSR is expressed in bitter receptor T2R-positive cells remains unresolved.

Our results suggest that *kokumi* substance-induced increases in $[Ca^{2+}]_i$ arise principally via mobilization from intracellular stores, because responses were essentially unaffected by depletion of extracellular Ca^{2+}. Conversely, responses were strongly abolished by pretreatment with U73122, demonstrating the involvement of phospholipase C in *kokumi* signal transduction. These results provide direct evidence of the existence of a functional *kokumi* receptor coupled to Ca^{2+} mobilization in taste cells. Generally, $[Ca^{2+}]_i$ response in taste cells seems to depend on two sources of Ca^{2+}. For depolarizing stimuli (KCl), Ca^{2+} influx is induced via activation of voltage-gated Ca^{2+} channels in presynaptic taste cells (DeFazio et al. 2006; Huang et al. 2005; Roberts et al. 2009). Interestingly, neither thapsigargin nor U73122 totally eliminated the Ca^{2+} responses evoked by *kokumi* stimuli. Whether this represents minor additional pathways remains unresolved.

In summary, our results demonstrate that *kokumi* substance-responsive cells are presynaptic and non-presynaptic taste cells. We observed that part of the *kokumi* substance-responsive cells did not respond to umami (MPG) or sweet (SC45647) stimuli. The *kokumi* flavor is defined as an enhancer of umami and sweet tastes. *Kokumi* substance-responding taste cells might involve enhancement of these basic tastes.

8.3 Structure-CaSR Activity Relation of *Kokumi* γ-Glutamyl Peptides

Abstract The previous studies indicated that *kokumi* γ-glutamyl peptides are perceived through the calcium-sensing receptor (CaSR) in human taste cells. Here we hypothesize that discovery of highly active CaSR agonist peptides may lead to the production of practical *kokumi* peptides. We analyzed the correlation between the structure and CaSR activity of γ-glutamyl peptides and found that the structural requirements for intense CaSR activity of γ-glutamyl peptides were as follows: the existence of an N-terminal γ-L-glutamyl residue; the presence of a moderately sized, aliphatic, neutral substituent at the second residue in an L-configuration; and the existence of a C-terminal carboxylic acid, preferably glycine, as the third constituent. The sensory analysis of γ-glutamyl peptides screened using CaSR activity assay and sensory evaluation showed that the norvaline-containing γ-glutamyl peptides, γ-glutamyl-norvalyl-glycine and γ-glutamyl-norvaline, were potent *kokumi* peptides than γ-glutamyl-valyl-glycine.

8.3.1 Introduction

The extracellular calcium-sensing receptor, CaSR, is a seven-transmembrane-spanning, G protein-coupled receptor (GPCR) that belongs to family C of the GPCR superfamily (Brown et al. 1993). This receptor is expressed on several tissues and organs, including the parathyroid gland and kidney, and plays a central role in extracellular calcium homeostasis in mammals (Chattopadhyay et al. 1997). Any increase in blood calcium level is sensed by CaSR, which suppresses the secretion of parathyroid hormone, stimulates the secretion of calcitonin, and induces the urinary excretion of calcium to reduce the level of blood calcium to normal range. CaSR is not only expressed in the parathyroid glands and kidney but is also found in many other organs, including the liver, heart, lung, gastrointestinal tract, pancreas, and central nervous system, suggesting its role in several biological functions (Brown and MacLeod 2001). CaSR has been reported to be activated by various substances such as cations (Ca^{2+}, Mg^{2+} and Gd^{3+}), basic peptides (protamine and polylysine), and polyamines (spermine) (Brown and MacLeod 2001). CaSR is expressed in a subpopulation of taste cells in mice and rats (Bystrova et al. 2010; San Gabriel et al. 2009) and is thought to exhibit a potential role in taste cell biology. Ninomiya et al. (1982) showed that a group of gustatory afferent nerve fibers responds to calcium and magnesium in mice. Several types of compounds have been reported to possess CaSR agonist activity. Amino acids are known to the large extracellular Venus flytrap domain (VFD), a structure common to all the members of class C GPCRs. CaSR is strongly activated by aromatic amino acids such as histidine, tryptophan, phenylalanine, and tyrosine but is weakly activated by other amino acids, such as arginine, lysine, valine, or glycine. All amino acids are thought to bind to the VFD-binding pocket through their amino and carboxyl groups (Conigrave et al. 2000; Conigrave and Hampson 2006). γ-Glutamyl peptides were reported to bind to the large extracellular VFD of CaSR (Wang et al. 2006; Broadhead et al. 2011).

As shown in the previous sections, it has been suggested that *kokumi* γ-glutamyl peptides are perceived through CaSR in human taste cells. *Kokumi* substances are known to modify the five basic tastes (especially sweet, salty, and umami) when added to basic tasting solutions or food although these substances have no taste themselves at the concentrations tested (Ueda et al. 1990; Ueda et al. 1997; Dunkel et al. 2007). Therefore, *kokumi* γ-glutamyl peptides may be identified via the sensory evaluation of the γ-glutamyl peptides selected by primary high-throughput screening using CaSR activity assay as an index. Thus, the search for active CaSR agonist peptides may lead to the discovery of other practical *kokumi* γ-glutamyl peptides.

In the present study, we clarified the structural requirement of γ-glutamyl peptides for the activation of the CaSR by dividing γ-Glu-Cys-Gly (glutathione, a representative *kokumi* substance) into three parts, namely, the N-terminus, second residue, and C-terminus. Each part was modified according to the previous studies on the structure–activity relation of γ-Glu-Cys-Gly (glutathione) (Cobb et al. 1982; De Craecker et al. 1997; Leslie et al. 2003). The modified γ-glutamyl peptides comprising non-proteinogenic amino acids, D-amino acids, and α-hydoxy acids were screened using CaSR assay as an index.

8.3.2 Materials and Methods

8.3.2.1 Materials

Chemicals for peptide synthesis were obtained from either Bachem AG (Switzerland), Watanabe Chemical Industries, Ltd. (Japan), or Kokusan Chemical Co., Ltd. (Japan). Commercially available peptide samples were purchased from Sigma-Aldrich (U.S.A.), Dojin Chemical Laboratory (Japan), Bachem AG, and Kokusan Chemical (Japan) and used as received. Dipeptide samples and tripeptides comprising proteinogenic amino acids were a contract manufactured product (Kokusan Chemical Co., Ltd. and Peptide Institute, Inc., Japan). Other peptides comprising non-proteinogenic amino acids and tetrapeptides were synthesized in our laboratories using liquid-phase synthesis and characterized by 1H-nuclear magnetic resonance (-NMR) spectroscopy (Bruker AVANCE400 (400 MHz)) and electrospray ionization (ESI)-mass spectrometry (-MS) (Thermo Quest TSQ700). Peptides were synthesized according to the methods reported by Amino et al. (2016).

8.3.2.2 Preparation of cRNA

Human cDNA of CaSR was previously described (Ohsu et al. 2010). Preparation of human CaSR cRNA was carried out from human kidney cDNA (Clontech) using a polymerase chain reaction (PCR) method. Primer oligonucleotide DNAs (forward primer N, 5′-ACT AAT ACG ACT CAC TAT AGG GAC CAT GGC ATT TTA TAG C TG CTG CTG G-3′, and reverse primer C, 5′-TTA TGA ATT CAC TAC GTT TTC T GT AAC AG-3′; National Center for Biotechnology Information [NCBI] accession number NM_000388) were synthesized, and PCR was performed using PfuUltra DNA Polymerase (Stratagene, USA). After reacting at a temperature of 94 °C for

3 min, a cycle of reactions at 94 °C for 30 s, 55 °C for 30 s, and 72 °C for 2 min was repeated 35 times, followed by a reaction step at 72 °C for 7 min. The plasmid vector pBR322 (TaKaRa, Japan) was digested with the restriction enzyme *Eco*RV and the PCR product was ligated to the EcoRV cleavage site of pBR322 using a ligation kit (Promega, USA). hCaSR cRNA was synthesized with a cRNA preparation kit (Ambion, USA) using the sequence as a template.

8.3.2.3 Determination of CaSR Activity Using *Xenopus* Oocytes

CaSR agonist-induced currents were characterized using the oocytes microinjected with hCaSR cRNA, as previously reported (Ohsu et al. 2010). Briefly, *Xenopus laevis* ovarian lobes were surgically removed, defolliculated, and treated with collagenase II. Approximately 10–20 ng of hCaSR cRNA was microinjected into oocytes, and the oocytes were incubated for 36–48 h at 15 °C in Barth's solution. Activation of CaSR (Gq class GPCR) expressed in the oocytes resulted in an increase in the intercellular calcium ion concentration, which concomitantly activated the endogenous calcium-dependent chloride channels with a measurable current. The oocytes were impaled by two electrodes in a voltage clamp configuration with a GeneClamp 500 (Axon) and responses were recorded using AxoScope 9.0 recording software (Axon) at a membrane potential of −70 mV. The oocytes were challenged with a 0.1–1000 μM solution of CaSR agonists in a perfusion buffer containing 96 mM sodium chloride (NaCl), 2 mM potassium chloride (KCl), 1 mM magnesium chloride ($MgCl_2$), 1.8 mM calcium chloride ($CaCl_2$), and 5 mM 4-(2-hydroxyethyl)-1-piperazineethansulfonic acid (HEPES) buffer (pH 7.2). The recorded peak current was indicative of the strength of receptor activation. Uninjected oocytes showed no response to calcium or the tested CaSR agonists.

8.3.2.4 Determination of CaSR Activity Using HEK-293 Cells

CaSR activity was measured using hCaSR-expressing HEK-293 cells according to the methods described in the previous section (Sect. 1).

8.3.2.5 Determination of Taste Recognition Threshold Concentrations of *Kokumi* Peptides

Taste recognition threshold concentrations of *kokumi* peptides were determined by sensory evaluations. The minimum peptide concentration at which the assessors recognize the increase in taste intensity as compared with umami and salty taste standard solution containing 0.05% monosodium glutamate (MSG), 0.05% disodium inosinate (IMP), and 0.5% NaCl without peptides was defined as the recognition threshold concentration in this study. Seven sample solutions at different concentrations were prepared as follows for each peptide in an umami and salty standard solution and subjected to a preliminary sensory evaluation: γ-Glu-Val-Gly, 0.05, 0.1, 0.2, 0.5, 1.0, 2.0, and 5.0 ppm; γ-Glu-Nva-Gly, 0.002, 0.005, 0.001, 0.02, 0.05, 0.1, and 0.2 ppm; and γ-Glu-Nva, 0.005, 0.01, 0.02, 0.05, 0.1, 0.2, and 0.5 ppm. A panel of five well-trained assessors (all were men; aged 24–52 years) tasted these samples in the given order and in the reverse order and compared them with a standard solution without peptides. The threshold value of the assessors was approximated by averaging the threshold values of individuals in three independent sessions. Human sensory analysis was conducted following the spirit of the Helsinki Declaration, and informed consent was obtained from all assessors. The experimental procedure was approved by the Ethics Board of the Institute of Food Sciences & Technologies, Ajinomoto (Tokyo, Japan).

8.3.3 *Results and Discussion*

8.3.3.1 CaSR Activities of γ-Glutamyl Dipeptides (γ-Glu-X)

We investigated CaSR activity of dipeptides and found that all the α-dipeptides were inactive and that most of the active dipeptides exhibited a "γ-glutamyl" structure (data not shown). The presence of a γ-peptidic bond between the N-terminal glutamic acid and the second residue was the most distinct structural feature of the active dipeptides. A common feature of class C GPCRs is the presence of an amino acid-binding site. As γ-glutamyl dipeptides have an α-amino acid structure, it may be reasonable to consider that some γ-glutamyl dipeptides may activate CaSR.

The CaSR activity of γ-glutamyl dipeptides is shown in Table 8.1. Among the γ-glutamyl dipeptides with proteinogenic amino acids, γ-Glu-Cys showed the most intense activity, while a relatively intense activity was found for γ-Glu-Val. On the contrary, some γ-glutamyl dipeptides, such as γ-Glu-Asp, γ-Glu-Asp, γ-Glu-Gln, γ-Glu-Glu, γ-Glu-Lys, γ-Glu-His, γ-Glu-Phe, and γ-Glu-Tyr, were inactive. The existence of a large side chain or a positively or negatively charged side chain abolished the observed activity. Of the γ-glutamyl dipeptides comprising non-proteinogenic amino acids, γ-Glu-Nva (Nva; norvaline), γ-Glu-Cle (Cle; cycloleucine), and γ-Glu-Abu (Abu; (S)-2-aminobutanoic acid) showed CaSR activities that were nearly the same as γ-Glu-Cys. These results indicate that the sulfhydryl group of the cysteine residue is not essential for CaSR activity; however, the presence of a small-to-medium-sized neutral alkyl side chain was necessary for high activity. γ-Glutamyl dipeptides containing D-amino acids (i.e., γ-Glu-D-Val) were inactive. In addition, neither α-Glu-Val nor γ-D-Glu-Val showed any activity, indicating that the γ-L-glutamyl linkage is required for CaSR activity. Replacement of the carboxyl group at the C-terminus with an amide group (γ-Glu-Val-NH_2), methoxycarbonyl group (γ-Glu-Val-OMe), or hydroxymethyl group (γ-Glu-Val-ol) resulted in the formation of compounds with very weak activities, suggesting that the carboxyl group in the third amino acid is important for CaSR activity.

Table 8.1 Determination of calcium-sensing receptor (CaSR) activity of γ-Glu-X peptides using *Xenopus* oocytes and HEK-293 cells

	CaSR activity (μM)	
	Xenopus oocyte[a]	HEK-293 cell[b]
γ-Glu-Cys	3	0.16–0.46
γ-Glu-Val	10	1.03
γ-Glu-Ala	50	1.24
γ-Glu-Thr	50	6.97
γ-Glu-Leu	100	5.07
γ-Glu-Ile	100	9.91
γ-Glu-Ser	300	11.0
γ-Glu-Orn	500	70.4
γ-Glu-Met	500	
γ-Glu-Asn		159
γ-Glu-Gly	1000	200
γ-Glu-Trp		300
γ-Glu-Pro		409
γ-Glu-Asp		Inactive
γ-Glu-Gln		Inactive
γ-Glu-Glu		Inactive
γ-Glu-Lys		Inactive
γ-Glu-His		Inactive
γ-Glu-Phe		Inactive
γ-Glu-Tyr		Inactive
γ-Glu-Cys(Me)	30	
γ-Glu-Tau	300	
γ-Glu-Cys(Me)(O)	300	
γ-Glu-Met(O)	1000	
γ-Glu-Nva		0.12
γ-Glu-Cle		0.15
γ-Glu-Abu		0.21
γ-Glu-Ape		0.49
γ-Glu-Cys(carboxymethyl)		1.69
γ-Glu-Tle		3.06
γ-Glu-Ser(Me)		3.19
γ-Glu-Nle		4.41
γ-Glu-β-homoAla		5.53
γ-Glu-allo-Ile		8.87
γ-Glu-Aib		15.4
γ-Glu-D-Val	Inactive	
γ-D-Glu-Val	Inactive	
α-Glu-Val	Inactive	
γ-Glu-Val-NH2	1000	

(continued)

Table 8.1 (continued)

	CaSR activity (μM)	
	Xenopus oocyte[a]	HEK-293 cell[b]
γ-Glu-Val-OMe	1000	
γ-Glu-Val-ol	1000	

Abbreviations: *Cys(Me)* S-(methyl)-cysteine, *Tau* taurine, *Cys(Me)(O)* S-(methyl)-cysteine sulfoxide, *Met(O)* methionine sulfoxide, *Nva* norvaline, *Cle* cycloleucine, *Abu* 2-aminobutanoic acid, *Ape* 3-aminopentanoic acid, *Cys(Carboxymethyl)* S-(Carboxymethyl)-cysteine, *Tle tert*-leucine, *Ser(Me)* O-methyl-serine, *Nle* norleucine, *β-homoAla* 3-aminobutanoic acid, *Aib* α-methylalanine, *Val-ol* valinol

[a]Minimum effective concentration

[b]EC_{50} values

8.3.3.2 CaSR Activities of γ-Glutamyl Tripeptides (γ-Glu-X-Gly)

CaSR activities of γ-Glu-X-Gly analogs with center position modification are shown in Table 8.2. The replacement of gamma-glutamyl residue cysteine side chain of the tripeptide with a large hydrophobic, basic, or acidic side chain resulted in their. As shown in Table 8.2, γ-Glu-Arg-Gly, γ-Glu-Asn-Gly, γ-Glu-Asp-Gly, γ-Glu-Gln-Gly, γ-Glu-Glu-Gly, γ-Glu-His-Gly, γ-Glu-Lys-Gly, γ-Glu-Met-Gly, γ-Glu-Orn-Gly, γ-Glu-Phe-Gly, and γ-Glu-Tyr-Gly were inactive. Tripeptides wherein the cysteine side chain was replaced by side chains or small-to-medium-sized hydrophobic side chains showed moderate or high CaSR activity. Incorporation of Val, containing a β-branched side chain, as the second residue (γ-Glu-Val-Gly) resulted in the production of an agonist with an EC_{50} value 30 times lower than that of γ-Glu-Cys-Gly, indicating that γ-Glu-Val-Gly is more potent as a CaSR agonist than γ-Glu-Cys-Gly. Furthermore, intense agonists were generated following the incorporation of non-proteinogenic L-amino acids with small-to-medium-sized hydrophobic side chains comprising one to three carbon(s) or two carbons and one sulfur or oxygen atom. As the activity of γ-Glu-Nva-Gly was approximately two orders of magnitude higher than that of γ-Glu-Nle-Gly (Nle; norleucine), the steric requirements for activity expression must be strictly controlled. However, γ-glutamyl tripeptides containing a D-amino acid residue (i.e., γ-Glu-D-Val-Gly) were inactive. These results suggest that the side chain of the second residue must carry an appropriate size and orientation for CaSR activation. In tripeptides, the Cys thiol group was thought to play an important role, but was not necessary for CaSR activity. Thus, the CaSR agonist function of γ-glutamyl tripeptides is independent of the functionality of the thiol group.

8.3.3.3 CaSR Activities of the N-Terminal- and C-Terminal-Modified Analogs of γ-Glu-Val-Y

CaSR activities of γ-Glu-Val-Y analogs with third position modification and other tri- and tetrapeptides are shown in Table 8.3. The replacement of the γ-glutamyl residue with a β-aspartyl residue (β-Asp-Val-Gly) or γ-D-glutamyl residue

Table 8.2 Determination of calcium-sensing receptor (CaSR) activity of γ-Glu-X-Gly peptides using *Xenopus* oocytes and HEK-293 cells

	CaSR activity	
	Xenopus oocyte[a]	HEK-293 cell[b]
γ-Glu-Val-Gly	0.1	0.030–0.075
γ-Glu-Cys-Gly	3	0.7
γ-Glu-Ala-Gly	10	0.016
γ-Glu-Ser-Gly	10	
γ-Glu-Gly-Gly	30	
γ-Glu-Ile-Gly	100	
γ-Glu-Thr-Gly	300	2.8
γ-Glu-Leu-Gly	300	
γ-Glu-Pro-Gly	300	
γ-Glu-Arg-Gly		Inactive
γ-Glu-Asn-Gly		Inactive
γ-Glu-Asp-Gly		Inactive
γ-Glu-Gln-Gly		Inactive
γ-Glu-Glu-Gly		Inactive
γ-Glu-His-Gly		Inactive
γ-Glu-Lys-Gly		Inactive
γ-Glu-Met-Gly		Inactive
γ-Glu-Orn-Gly		Inactive
γ-Glu-Phe-Gly		Inactive
γ-Glu-Tyr-Gly		Inactive
γ-Glu-Cys(Me)-Gly	3	
γ-Glu-Abu-Gly	3	0.025
γ-Glu-Algly-Gly		0.011
γ-Glu-Ser(Me)-Gly		0.037
γ-Glu-Nva-Gly		0.052–0.055
γ-Glu-Tle-Gly		0.043–0.090
γ-Glu-Pen-Gly		0.16–0.23
γ-Glu-Aib-Gly		0.35–1.17
γ-Glu-Cle-Gly		0.5–1.5
γ-Glu-*allo*-Ile-Gly		0.63
γ-Glu-*allo*-Thr-Gly		0.7
γ-Glu-Hse-Gly		1.18
γ-Glu-Cys(n-butyl)-Gly		4
γ-Glu-Nle-Gly		5
γ-Glu-Cys(1,2-dicarboxyethyl)-Gly		5–10
γ-Glu-Cys(allyl)-Gly	100	
γ-Glu-Cys(n-propyl)-Gly		117

Abbreviations: *Algly* α-allylglycine, *Pen* penicillamine, *Hse* homoserine, *Cys(n-butyl)* S-(n-butyl)-cysteine, *Cys(allyl)*, S-(allyl)-cysteine, *Cys(n-propyl)* S-(n-propyl)-cysteine

[a]Minimum effective concentration

[b]EC_{50} values

Table 8.3 Determination of calcium-sensing receptor (CaSR) activity of γ-Glu-Val-Y peptides and other tri- or tetrapeptides using *Xenopus* oocytes and HEK-293 cells

	CaSR activity	
	Xenopus oocyte[a]	HEK-293 cell[b]
γ-Glu-Val-Gly	0.1	0.030–0.075
γ-Glu-Val-Gln	10	0.3–0.5
γ-Glu-Val-Cys	10	3.7
γ-Glu-Val-Pro	30	
γ-Glu-Val-Ser	30	
γ-Glu-Val-Phe	30	
γ-Glu-Val-Asn	30	2.8
γ-Glu-Val-Orn	100	
γ-Glu-Val-His	100	
γ-Glu-Val-Ala	300	
γ-Glu-Val-Thr	1000	
γ-Glu-Val-Met	1000	
γ-Glu-Val-Asp	1000	
γ-Glu-Val-Arg	1000	
γ-Glu-Val-Lys	1000	
γ-Glu-Val-Glu	1000	
γ-Glu-Val-Val	1000	
γ-Glu-Val-D-Ala		1.26
γ-Glu-Val-D-Ser		4.27
γ-Glu-Ala-Leu		4.17
γ-Glu-Abu-Leu		2.49
γ-Glu-Abu-Pro		4.59
γ-Glu-D-Val-Gly		Inactive
γ-D-Glu-Val-Gly		Inactive
γ-D-Glu-D-Val-Gly	100	Inactive
α-Glu-Val-Gly		>10
α-D-Glu-Val-Gly		Inactive
α-Glu-D-Val-Gly		Inactive
α-D-Glu-D-Val-Gly		Inactive
β-Asp-Val-Gly		Inactive
β-Asp-Val-β-Ala		Inactive
γ-Glu-Val-β-Ala		Inactive
γ-Glu-Cys-β-Ala		5–10
γ-Glu-Abu-β-Ala		2.45
γ-Glu-Ala-β-Ala		9.96
γ-Glu-Val-Gly-Gly		4.91
γ-Glu-Val-Gly-Gln		2.03
γ-Glu-Val-glycolic acid		0.05
γ-Glu-Abu-glycolic acid		0.14
γ-Glu-Tle-glycolic acid		0.065–0.079
γ-Glu-Val-L-lactic acid		0.102
γ-Glu-Abu-L-lactic acid		0.038
γ-Glu-Tle-L-lactic acid		0.057

[a]Minimum effective concentration
[b]EC_{50} values

(γ-D-Glu-Val-Gly) resulted in the formation of inactive tripeptides. The α-glutamyl peptide α-Glu-Val-Gly showed weak activity and almost all the C-terminal modified analogs of γ-Glu-Val-Gly comprising a proteinogenic amino acid were active. However, all tripeptide C-terminal-modified analogs were less active than γ-Glu-Val-Gly. The replacement of Gly with Gln and Cys yielded tripeptides (γ-Glu-Val-Gln and γ-Glu-Val-Cys) that were active at concentrations three orders of magnitude less than that of γ-Glu-Val-Gly. The incorporation of β-Ala (β-alanine) or D-amino acids, such as D-Ala and D-Ser, instead of Gly (γ-Glu-Val-β-Ala, γ-Glu-Cys-β-Ala, γ-Glu-Abu-β-Ala, γ-Glu-Ala-β-Ala, γ-Glu-Val-D-Ala, and γ-Glu-Val-D-Ser) resulted in the production of inactive or weak peptides. The replacement of Gly with an α-hydroxy acid, such as glycolic acid or L-lactic acid (γ-Glu-Val-Glycolic acid, γ-Glu-Abu-Glycolic acid, γ-Glu-Tle-Glycolic acid, γ-Glu-Val-L-lactic acid, γ-Glu-Abu-L-lactic acid, and γ-Glu-Tle-L-lactic acid), resulted in intense CaSR activities similar to those reported with γ-Glu-Val-Gly. Among the tetrapeptides carrying another amino acid to the C-terminus of γ-Glu-Val-Gly, γ-Glu-Val-Gly-Gly and γ-Glu-Val-Gly-Gln showed weak activities.

8.3.3.4 Summary of the CaSR Activities of γ-Glutamyl Peptides

The CaSR activities of several dipeptide and tripeptide analogs indicated the importance of the γ-glutamyl residue to mediate peptide activity, with a few exceptions. The exact alignment as well as the presence of the two charged groups on the N-terminal residue are essential to mediate binding to CaSR. A small-to-medium-sized side chain on the second residue in L-configuration, as observed in Val, Abu, Nva, α-allylglycine (Algly), O-(methyl)-serine (Ser(Me)), and tert-leucine (Tle), most likely enhanced the binding of the peptides to the receptor site in an active orientation. The presence of a third residue, especially one with a free carboxy terminus and no side chain (i.e., Gly), significantly enhanced the activity. However, the existence of this C-terminal third residue is preferable but not essential for the activity. These findings were consistent with the reports by Wang et al. (2006) and Broadhead et al. (2011), wherein the importance of the interaction between both N-terminal amino acid and the C-terminal carboxyl groups of γ-glutamyl tripeptides and the active site of CaSR was discussed.

Fig. 8.12 shows the putative interaction between γ-Glu-Val-Gly and CaSR, wherein VFD mediates γ-Glu-Val-Gly. The following three interactions must occur between γ-glutamyl peptides and the VFD of CaSR: γ-L-glutamyl residue with a zwitterionic binding site, which has an amino acid binding site for VFD; hydrophobic side chain of Val with a hydrophobic interaction site; and C-terminal carboxylic acid of Gly with an ionic binding site.

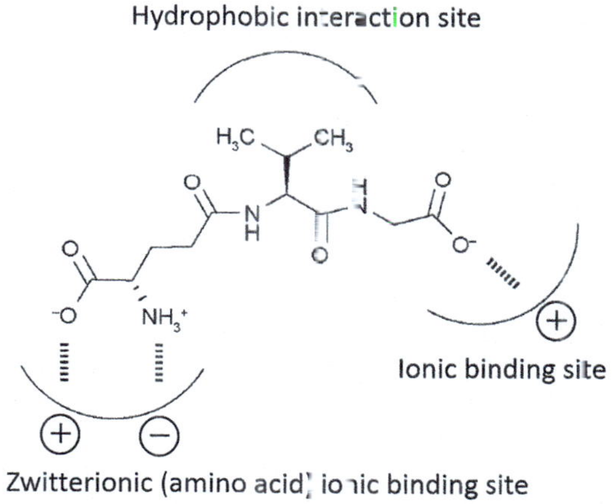

Fig. 8.12 Putative interaction between γ-Glu-Val-Gly and CaSR

Table 8.4 Taste recognition threshold concentration of γ-Glu-Val-Gly, γ-Glu-Nva-Gly, and γ-Glu-Nva in umami and salty solution[a]

	Taste recognition threshold concentration	
	Weight conc. (ppm)	Molar conc. (μM)
γ-Glu-Val-Gly	0.47	1.55
γ-Glu-Nva-Gly	0.028	0.092
γ-Glu-Nva	0.077	0.31

[a]Umami and salty solution contains 0.05% MSG, 0.05% IMP, and 0.5% NaCl

8.3.3.5 Sensory Activity of *Kokumi* γ-Glutamyl Peptides

Sensory analysis was conducted for γ-glutamyl peptides screened based on CaSR activity assay. EC_{50} values of γ-Glu-Val-Gly, γ-Glu-Nva-Gly, and γ-Glu-Nva in CaSR assay were 0.030–0.075, 0.052–0.055, and 0.12 μM, respectively, whereas the value reported for γ-Glu-Cys-Gly (glutathione) was 0.7 μM.

The results of sensory evaluation are summarized in Table 8.4. The taste recognition threshold concentration of γ-Glu-Nva-Gly was determined as 0.028 ppm (0.092 μM), and that of γ-Glu-Nva was 0.077 ppm (0.31 μM). These values were significantly lower than the taste recognition threshold concentration of γ-Glu-Val-Gly (0.47 ppm; 1.55 μM). Although Nva is a non-proteinogenic amino acid, γ-Glu-Nva-Gly was recently detected in human serum samples (Hirayama et al. 2014).

The reason underlying the intense sensory activity for γ-Glu-Nva-Gly and γ-Glu-Nva relative to the activity for γ-Glu-Val-Gly is incompletely understood. Quantitative analysis of the sensory activity of these peptides is currently under investigation and will be published elsewhere.

8.3.4 Conclusion

The correlation between the structure and CaSR activity of γ-glutamyl peptides was investigated, and the structural requirements for intense CaSR activity of γ-glutamyl peptides were determined as follows: existence of an N-terminal γ-L-glutamyl residue; existence of a moderately sized, aliphatic, neutral substituent at the second residue in an L-configuration; and existence of a C-terminal carboxylic acid, preferably with Gly as the third constituent. The sensory analysis of γ-glutamyl peptides screened based on CaSR activity assay and sensory evaluation revealed γ-Glu-Nva-Gly and γ-Glu-Nva to be more potent *kokumi* peptides than γ-Glu-Val-Gly.

Acknowledgments We sincerely thank Dr. Kiyoshi Miwa, Dr. Tohru Kouda, Dr. Toshihisa Kato, Naohiro Miyamura, and Dr. Yuzuru Eto of Ajinomoto Co., Inc., for their encouragement and continued support of this work. We also thank Dr. Ken Iwatsuki, Dr. Hisayuki Uneyama, and Dr. Kunio Torii for providing the antibodies and for valuable suggestions. We thank Takeaki Ohsu, Sen Takeshita, Reiko Yasuda, and Fumie Futaki for their technical assistance on CaSR activity assay. We thank Dr. Yusuke Amino, Dr. Masakazu Nakazawa, Yuki Tahara, Megumi Kaneko, and Toshihiro Hatanaka for synthesis of peptides. We are grateful to Hiroaki Nagasaki, Tomohiko Yamanaka, Fusataka Kenmotsu, Takashi Miyaki, and Takaho Tajima for conducting sensory evaluation. We are also grateful to the panelists who have participated on the sensory evaluation.

References

Amino Y, Nakazawa M, Kaneko M, Miyaki T, Miyamura N, Maruyama Y, Eto Y (2016) Structure-CaSR activity relation of *kokumi* γ-glutamyl peptides. Chem Pharm Bull 64:1181–1189

Bleasdale JE, Thakur NR, Gremban RS, Bundy GL, Fitzpatrick FA et al (1990) Selective inhibition of receptor-coupled phospholipase C-dependent processes in human platelets and polymorphonuclear neutrophils. J Pharmacol Exp Ther 255:756–768

Broadhead GK, Mun HC, Avlani VA, Jourdan O, Church WB, Christopoulos A, Delbridge L, Conigrave AD (2011) Allosteric modulation of the calcium-sensing receptor by gamma-glutamyl peptides: inhibition of PTH secretion, suppression of intracellular cAMP levels, and a common mechanism of action with L-amino acids. J Biol Chem 286:8786–8797

Brown EM, MacLeod RJ (2001) Extracellular calcium sensing and extracellular calcium signaling. Physiol Rev 81:239–297

Brown EM, Gamba G, Riccardi D, Lombardi M, Butters R et al (1993) Cloning and characterization of an extracellular Ca2+−sensing receptor from bovine parathyroid. Nature 366:575–580. EM

Bystrova MF, Romanov RA, Rogachevskaja OA, Churbanov GD, Kolesnikov SS (2010) Functional expression of the extracellular-Ca2+−sensing receptor in mouse taste cells. J Cell Sci 123:972–982

Caicedo A, Jafri MS, Roper SD (2000) In situ Ca2+ imaging reveals neurotransmitter receptors for glutamate in taste receptor cells. J Neurosci 20:7978–7985

Caicedo A, Kim KN, Roper SD (2002) Individual mouse taste cells respond to multiple chemical stimuli. J Physiol 544:501–509

Chattopadhyay N, Vassilev PM, Brown EM (1997) Calcium-sensing receptor: roles in and beyond systemic calcium homeostasis. Biol Chem 378:759–768

Chaudhari N, Roper SD (2010) The cell biology of taste. J Cell Biol 190:285–296

Cobb MH, Heagy W, Danner J, Lenhoff HM, Marshall GR (1982) Structural and conformational properties of peptides interacting with the glutathione receptor of hydra. Mol Pharmacol 21:629–631

Conigrave AD, Hampson DR (2006) Broad-spectrum L-amino acid sensing by class 3 G-protein-coupled receptors. Trends Endocrinol Metab 17:398–407

Conigrave AD, Quinn SJ, Brown EM (2000) L-amino acid sensing by the extracellular Ca^{2+} – sensing receptor. Proc Natl Acad Sci U S A 97:4814–4819

Dando R, Roper SD (2009) Cell-to-cell communication in intact taste buds through ATP signalling from pannexin 1 gap junction hemichannels. J Physiol 587:5899–5906

De Craecker S, Verbruggen C, Rajan PK, Smith K, Haemers A, Fairlamb A (1997) Characterization of the peptide substrate specificity of glutathionylspermidine synthetase from *Crithidia fasciculata*. Mol Biochem Parasitol 84:25–32

DeFazio RA, Dvoryanchikov G, Maruyama Y, Kim JW, Pereira E et al (2006) Separate populations of receptor cells and presynaptic cells in mouse taste buds. J Neurosci 26 3971–3980

Dunkel A, Koster J, Hofmann T (2007) Molecular and sensory characterization of gamma-glutamyl peptides as key contributors to the kokumi taste of edible beans (Phaseolus vulgaris L.). J Agric Food Chem 55:6712–6719

Dvoryanchikov G, Tomchik SM, Chaudhari N (2007) Biogenic amine synthesis and uptake in rodent taste buds. J Comp Neurol 505:302–313

Finger TE, Danilova V, Barrows J, Bartel DL, Vigers AJ et al (2005) ATP signaling is crucial for communication from taste buds to gustatory nerves. Science 310:1495–1499

Gowen M, Stroup GB, Dodds RA, James IE, Votta BJ et al (2000) Antagonizing the parathyroid calcium receptor stimulates parathyroid hormone secretion and bone formation in osteopenic rats. J Clin Invest 105:1595–1604

Helmchen G (2000) In: Yuste R, Lanni F, Konnerth A (eds) Calibration of fluorescent calcium indicators. Cold Spring Harbor Laboratory, Cold Spring Harbor

Hirayama A, Igarashi K, Tomita M, Soga T (2014) Development of quantitative method for determination of γ-glutamyl peptides by capillary electrophoresis tandem mass spectrometry: an efficient approach avoiding matrix effect. J Chromatogr 1369:161–169

Huang L, Shanker YG, Dubauskaite J, Zheng JZ, Yan W et al (1999) Ggamma13 colocalizes with gustducin in taste receptor cells and mediates IP3 responses to bitter denatonium. Nat Neurosci 2:1055–1062

Huang YJ, Maruyama Y, Lu KS, Pereira E, Plonsky I et al (2005) Mouse taste buds use serotonin as a neurotransmitter. J Neurosci 25:843–847

Huang AL, Chen X, Hoon MA, Chandrashekar J, Guo W et al (2006) The cells and logic for mammalian sour taste detection. Nature 442:934–938

Huang YJ, Maruyama Y, Dvoryanchikov G, Pereira E, Chaudhari N et al (2007) The role of pannexin 1 hemichannels in ATP release and cell-cell communication in mouse taste buds. Proc Natl Acad Sci U S A 104:6436–6441

Huang YA, Maruyama Y, Stimac R, Roper SD (2008) Presynaptic (type III) cells in mouse taste buds sense sour (acid) taste. J Physiol 586:2903–2912

Iwatsuki K, Ichikawa R, Hiasa M, Moriyama Y, Torii K et al (2009) Identification of the vesicular nucleotide transporter (VNUT) in taste cells. Biochem Biophys Res Commun 388:1–5

Kinnamon JC, Taylor BJ, Delay RJ, Roper SD (1985) Ultrastructure of mouse vallate taste buds. I. Taste cells and their associated synapses. J Comp Neurol 235:48–60

Leslie EM, Bowers RJ, Deeley RG, Cole SPC (2003) Structural requirements for functional interaction of glutathione tripeptide analogs with the human multidrug resistance protein 1 (MRP1). J Pharmacol Exp Ther 304:643–653

Li X, Staszewski L, Xu H, Durick K, Zoller M et al (2002) Human receptors for sweet and umami taste. Proc Natl Acad Sci U S A 99:4692–4696

Maruyama Y, Pereira E, Margolskee RF, Chaudhari N, Roper SD (2006) Umami responses in mouse taste cells indicate more than one receptor. J Neurosci 26:2227–2234

Maruyama Y, Yasuda R, Kuroda M, Eto Y (2012) Kokumi substances, enhancers of basic tastes, induce responses in calcium-sensing receptor expressing taste cells. PLoS One 7:e34489
McCaughey SA, Forestell CA, Tordoff MG (2005) Calcium deprivation increases the palatability of calcium solutions in rats. Physiol Behav 84:335–342
Medler KF, Margolskee RF, Kinnamon SC (2003) Electrophysiological characterization of voltage-gated currents in defined taste cell types of mice. J Neurosci 23:2608–2617
Michlig S, Damak S, Le Coutre J (2007) Claudin-based permeability barriers in taste buds. J Comp Neurol 502:1003–1011
Montmayeur JP, Liberles SD, Matsunami H, Buck LB (2001) A candidate taste receptor gene near a sweet taste locus. Nat Neurosci 4:492–498
Murray RG (1993) Cellular relations in mouse circumvallate taste buds. Microsc Res Tech 26:209–224
Nelson G, Hoon MA, Chandrashekar J, Zhang Y, Ryba NJ et al (2001) Mammalian sweet taste receptors. Cell 106:381–390
Nelson G, Chandrashekar J, Hoon MA, Feng L, Zhao G et al (2002) An amino-acid taste receptor. Nature 416:199–202
Ninomiya Y, Tonosaki K, Funakoshi M (1982) Gustatory neural response in the mouse. Brain Res 244:370–373
Nofre G, Tinti JM, Chatzopoulos FO (1987) Sweetening agents. US patent 4921939
Ogura T, Kinnamon SC (1999) IP(3)-independent release of Ca2+ from intracellular stores: a novel mechanism for transduction of bitter stimuli. J Neurophysiol 82:2657–2666
Ohsu T, Amino Y, Nagasaki H, Yamanaka T, Takeshita S et al (2010) Involvement of the calcium-sensing receptor in human taste perception. J Biol Chem 285:1016–1022
Richter TA, Dvoryanchikov GA, Chaudhari N, Roper SD (2004) Acid-sensitive two-pore domain potassium (K2P) channels in mouse taste buds. J Neurophysiol 92:1928–1936
Roberts CD, Dvoryanchikov G, Roper SD, Chaudhari N (2009) Interaction between the second messengers cAMP and Ca2+ in mouse presynaptic taste cells. J Physiol 587:1657–1668
Rodriguez M, Nemeth E, Martin D (2005) The calcium-sensing receptor: a key factor in the pathogenesis of secondary hyperparathyroidism. Am J Physiol Renal Physiol 288:F253–F264
Romanov RA, Rogachevskaja OA, Bystrova MF, Jiang P, Margolskee RF et al (2007) Afferent neurotransmission mediated by hemichannels in mammalian taste cells. EMBO J 26:657–667
Rossler P, Kroner C, Freitag J, Noe J, Breer H (1998) Identification of a phospholipase C beta subtype in rat taste cells. Eur J Cell Biol 77:253–261
Rybczynska A, Lehmann A, Jurska-Jasko A, Boblewski K, Orlewska C et al (2006) Hypertensive effect of calcilytic NPS 2143 administration in rats. J Endocrinol 191:189–195
Salari H, Bramley A, Langlands J, Howard S, Chan-Yeung M et al (1993) Effect of phospholipase C inhibitor U-73122 on antigen-induced airway smooth muscle contraction in Guinea pigs. Am J Respir Cell Mol Biol 9:405–410
San Gabriel A, Uneyama H, Maekawa T, Torii K (2009) The calcium-sensing receptor in taste tissue. Biochem Biophys Res Commun 378:414–418
Thompson AK, Mostafapour SP, Denlinger LC, Bleasdale JE, Fisher SK (1991) The aminosteroid U-73122 inhibits muscarinic receptor sequestration and phosphoinositide hydrolysis in SK-N-SH neuroblastoma cells. A role for Gp in receptor compartmentation. J Biol Chem 266:23856–23862
Toelstede S, Hofmann T (2009) Kokumi-active glutamyl peptides in cheeses and their biogeneration by Penicillium roquefortii. J Agric Food Chem 57:3738–3748
Tomchik SM, Berg S, Kim JW, Chaudhari N, Roper SD (2007) Breadth of tuning and taste coding in mammalian taste buds. J Neurosci 27:10840–10848
Tordoff MG, Shao H, Alarcon LK, Margolskee RF, Mosinger B et al (2008) Involvement of T1R3 in calcium-magnesium taste. Physiol Genomics 34:338–348
Ueda Y, Sakaguchi M, Hirayama K, Miyajima R, Kimizuka A (1990) Characteristic flavor constituents in water extract of garlic. Agric Biol Chem 54:163–169

Ueda Y, Tsubuku T, Miyajima R (1994) Composition of sulfur-containing components in onion and their flavor characters. Biosci Biotechnol Biochem 58:108–110
Ueda T, Yonemitsu M, Tsubuku T, Sakaguchi M, Miyajima R (1997) Flavor characteristics of glutathione in raw and cooked foodstuffs. Biosci Biotechnol Biochem 61:1977–1980
Wang M, Yao Y, Kuang D, Hampson DR (2006) Activation of family C G-protein-coupled receptors by the tripeptide glutathione. J Biol Chem 281:8864–8870
Yee CL, Yang R, Bottger B, Finger TE, Kinnamon JC (2001) "Type III" cells of rat taste buds: immunohistochemical and ultrastructural studies of neuron-specific enolase, protein gene product 9.5, and serotonin. J Comp Neurol 440:97–108
Zhang Y, Hoon MA, Chandrashekar J, Mueller KL, Cook B et al (2003) Coding of sweet, bitter, and umami tastes: different receptor cells sharing similar signaling pathways. Cell 112:293–301

Chapter 9
Mouse Trigeminal Neurons Respond to *Kokumi* Substances

Sara C. M. Leijon, Nirupa Chaudhari, and Stephen D. Roper

Abstract Using in vivo confocal Ca^{2+} imaging, we investigated whether oral application of *kokumi* substances elicits responses in trigeminal somatosensory ganglion neurons of the mouse. Our results show that 100 μM γEVG (γ-Glu-Val-Gly), a potent *kokumi* stimulus, evokes responses in a very small fraction (0.6%) of neurons in area V3 (oral sensory field) of the trigeminal ganglion. By comparison, cooled artificial saliva elicited thermal-evoked responses in >7% of V3 trigeminal ganglion neurons. γEVG-evoked responses were small and quite variable, with latencies ranging from 2 to over 200 s. Co-application of the calcium-sensing receptor (CaSR) inhibitor NPS-2143 significantly decreased γEVG-evoked activity. Furthermore, we show that four additional *kokumi* substances evoked responses in mouse trigeminal ganglion neurons. All neurons responding to *kokumi* compounds were small cells, with mean diameters below 20 μm. In summary, our data show that certain physiological and pharmacological properties of responses to *kokumi* compounds can be recorded from sensory neurons in the trigeminal ganglion of living mice. Thus, sensory neurons in the somatosensory trigeminal ganglia may transmit signals from the oral cavity to the central nervous system to generate the texture perceptions that are part of the enigmatic sensations evoked by *kokumi* substances.

Keywords Trigeminal ganglion · Somatosensations · Texture · Viscosity · Calcium imaging · Orosensations

---

S. C. M. Leijon
Department of Physiology & Biophysics, Miller School of Medicine, University of Miami, Miami, FL, USA

N. Chaudhari · S. D. Roper (✉)
Department of Physiology & Biophysics, Miller School of Medicine, University of Miami, Miami, FL, USA

Department of Otolaryngology, Miller School of Medicine, University of Miami, Miami, FL, USA

Neuroscience Graduate Program, University of Miami, Miami, FL, USA
e-mail: sroper@med.miami.edu

T. Nishimura, M. Kuroda (eds.), *Koku in Food Science and Physiology*,
https://doi.org/10.1007/978-981-13-8453-0_9

9.1 Introduction

Sensations elicited by *kokumi* substances are difficult to define and to quantify. It is often described as "mouthfulness," "thickness," and "continuity (lingeringness)" (Dunkel et al. 2007)—characteristics that cannot readily be explained in terms of taste alone. *Kokumi* compounds, though themselves evoking no taste, are also claimed to enhance sweet, salty, and umami tastes (Ohsu et al. 2010; but see Dunkel et al. 2007). These features are not easily measured and often rely on human sensory panels and verbal descriptions. Finding an animal model that can be used to test and quantify *kokumi* sensations would be valuable for research and development of *kokumi* compounds. This chapter outlines progress along those lines and describes a new approach to analyzing *koku* taste in a mouse model.

Afferent somatosensory neurons in the trigeminal ganglion innervate the tongue and oral surfaces. These neurons are sensitive to many different modalities, including mechanical, thermal, chemesthetic, and nociceptive stimuli. Trigeminal sensory neurons may thus participate in transmitting sensations such as texture and viscosity, which could be perceived as mouthfulness, thickness, and continuity. Interestingly, trigeminal sensory neurons in the rat are reported to express a calcium-sensing receptor, CaSR (Heyeraas et al. 2008), which has been shown to be activated by *kokumi* substances (Ohsu et al. 2010; Amino et al. 2016). Collectively, these features provide a rationale for studying the involvement of trigeminal ganglion somatosensory neurons in responses to *kokumi* compounds. Accordingly, we have taken advantage of a powerful new experimental approach that combines Ca^{2+} imaging, genetically engineered mice, and in vivo scanning laser confocal microscopy (Fig. 9.1) to investigate responses of mouse trigeminal ganglion neurons to oral stimulation with *kokumi* stimuli. Our aim was to determine whether trigeminal gan-

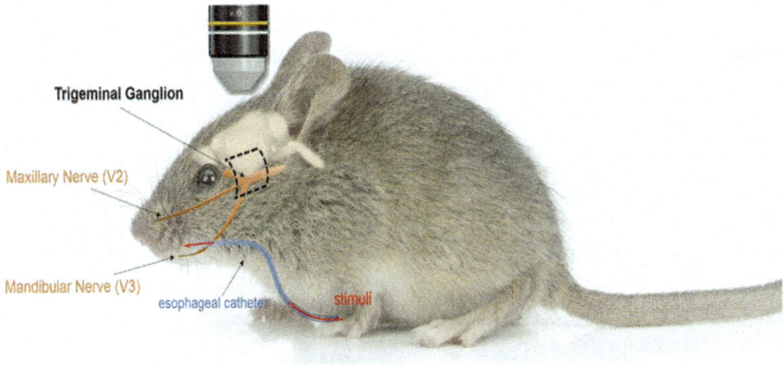

Fig. 9.1 Schematic of the trigeminal ganglion location in the mouse. The trigeminal ganglion (orange) resides at the bottom of the cranium, below the brain. The mandibular and maxillary branches of the trigeminal nerve innervate the oral cavity. A surgery is performed to carefully expose the trigeminal ganglion via a dorsolateral window (dashed rectangle). The surgically prepared mouse is thereafter transferred to a confocal laser scanning microscope, and neuronal calcium responses are recorded with a 10x-long working distance objective

glion neurons respond to *kokumi* substances that are presented in the oral cavity and, in particular, to stimulation by the prototypic *koku* substance γ-glutamyl-valinyl-glycine (γEVG). Using this animal model, we tested sensory neuron responses to γEVG in the living animal and characterized some of the properties of this *kokumi* compound. Our particular interests were in the viscous sensations elicited by γEVG and in the identity of possible membrane receptors for that compound.

9.2 Methodology

9.2.1 Animal Model

We used transgenic mice that express the calcium indicator GCaMP3 or GCaMP6s in sensory neurons, including those in the trigeminal ganglion (Leijon et al. 2019). GCaMP3 mice were obtained from X. Dong, Johns Hopkins, and express GCaMP3 as a knock-in/knockout at the Pirt locus (Kim et al. 2014). We also used floxed GCaMP6s (Jax #024106) crossed with Pirt-Cre mice (X Dong, Johns Hopkins) to generate mice in which sensory neurons expressed GCaMP6s. All mice were back-crossed to C57Bl/6 for 8–10 generations and adult mice of both sexes were used. Animals were housed with a 12-hour light cycle and food and water *ad libitum* All experiments were carried out during the daylight cycle. All procedures for surgery and euthanasia were reviewed and approved by the University of Miami IACUC.

9.2.2 Surgical Preparation

Mice were anesthetized with ketamine and xylazine (intraperitoneally 120 mg/kg ketamine, 10 mg/kg xylazine), and anesthetic depth was monitored by the hind paw withdrawal reflex. Ketamine booster injections were administered to ensure continued surgical plane of anesthesia throughout the surgery and imaging session. The animal's core temperature was continuously monitored with a rectal probe and maintained between 35 and 36 °C.

For the initial stage of surgery, the anesthetized mouse was placed supine on a far-infrared surgical warming pad (DCT-15, Kent Scientific). The trachea was exposed and cannulated for facilitated respiration during oral stimulus presentation. A flexible tube was inserted through the esophagus to produce a uniform stimulus delivery into the oral cavity (Sollars and Hill 2005; Wu, Dvoryanchikov et al. 2015; Leijon et al. 2019). Next, the mouse was placed in prone position for surgical exposure of the region of the trigeminal ganglion that innervates the oral cavity, V3 (Fig. 9.2). Muscle tissue on the dorsolateral side of the skull was cauterized and the zygomatic arch removed. A small cranial window was drilled open and hemispherectomy was performed by careful aspiration to allow optical access to the dorsal part of the trigeminal ganglion. The head was stabilized by affixing a custom

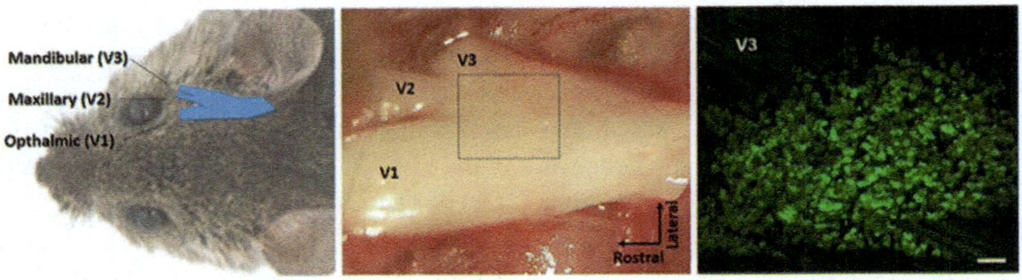

Fig. 9.2 Location and surgical view of the trigeminal ganglion. (**a**) Dorsal view, displaying the approximate position of the trigeminal ganglion (blue). (**b**) Photograph showing the V3 region of the surgically exposed trigeminal ganglion and the area where functional imaging was carried out (rectangle). (**c**) Under the confocal laser scanning microscope, the calcium indicator, GCaMP, in trigeminal ganglion neurons fluoresces green. Stimulus-evoked responses generate robust fluorescence increases in stimulated neurons. A typical recording field contains around 1000 ganglion neurons and can be imaged for over 5 h. Scale bar, 100 μm

head holder to the skull with a nylon screw and dental acrylic. To maintain a favorable neuronal environment, from the moment of exposure the ganglion was intermittently rinsed with Tyrode's solution at 35 °C.

9.2.3 *Functional In Vivo Confocal Calcium Imaging*

The surgically prepared mouse was transferred to the stage of an Olympus FV1000 confocal microscope equipped with a 10x-long working distance objective (UPlanFl, N.A. 0.3). Confocal scans of GCaMP3-/GCAMP6s-labeled ganglion neurons were taken at ~1 hz using 488-nm laser excitation with a 505–605-nm emitter filter. Oral stimuli were applied as 1 mL over 10 s, delivered retrogradely into the buccal cavity via the esophageal tube. When temperature stimuli were applied, the resulting oral temperature was measured with a miniature thermal probe (RET-4, Physitemp Instruments, LLC) inserted into the mouth and connected to a wireless transmitter (UWBT-TC-UST-NA, OMEGA Engineering, Inc.).

9.2.4 *Reagents*

All reagents were purchased from Sigma apart from *kokumi* substances, which were obtained from Ajinomoto Co., Inc. These included glutathione (GSH; CAS 70-18-8), γ-glutamyl-alpha-aminobutyrate (γ-Glu-Abu; CAS 16869-42-4), γ-glutamyl-valinyl-glycine (γEVG, γ-Glu-Val-Gly; CAS 38837-70-6), poly-L-lysine (CAS 28211-04-3), cinacalcet-HCl (CAS 364782-34-3), and the calcium-sensing receptor inhibitor NPS-2143. Stock solutions of *kokumi* substances were dissolved in H_2O

and stored at –20°C. On the experimental day, *kokumi* stimuli were diluted to working solutions in artificial saliva containing 15 mM NaCl, 22 mM KCl, 3 mM $CaCl_2$, 0.6 mM $MgCl_2$, pH 5.8.

9.2.5 Data Analysis

Optical scans of neuronal fluorescence were digitized using Fluoview software (Olympus) and video images were stabilized using FIJI (ImageJ). Baseline-subtracted videos were used to identify areas with potential neuronal responses and regions of interest (ROIs) were manually drawn over individual ganglion neurons. ROIs were analyzed with MatLab (MathWorks) using custom code, modified from Wu et al. (2015), to correct for any small baseline drifts. Calcium transients were quantified as peak stimulus-evoked fluorescence change (ΔF) divided by baseline fluorescence (F_0), i.e., $\Delta F/F_0$. The criterion for identifying responses was $\Delta F/F_0 > 5$ standard deviations above baseline (Fig. 9.3). For thermal stimuli, criteria also included that responses occurred at consistent latencies after stimulus onset. A single field of view, within the V3 area of the trigeminal ganglion, was analyzed for each experimental animal. At the end of an experiment, the animal was euthanized by CO_2 followed by cervical dislocation.

Unless otherwise specified, statistical significance was determined using two-tailed Student's t-test between two groups, otherwise by unpaired or paired one-way ANOVA, using Prism v.6 (GraphPad). Statistical significance was defined as $p < 0.05$.

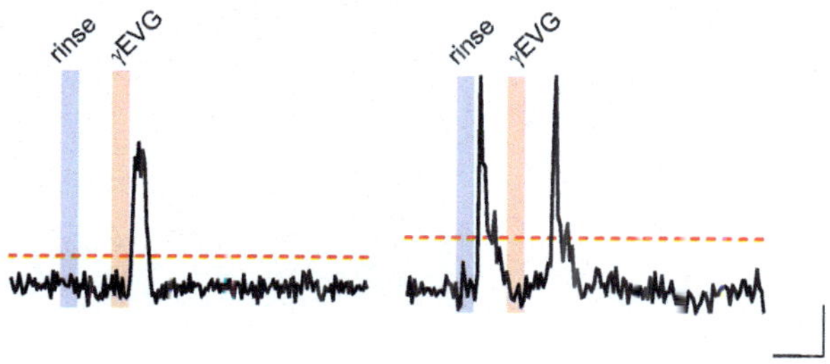

Fig. 9.3 Examples of responses ($\Delta F/F_0$) from two separate neurons to oral stimulation with 100 μM γEVG. Here, as in later figures, stimuli were superfused into the oral cavity for 10 s (shaded areas). Red dashed lines mark criterion level above which $\Delta F/F_0$ must rise to be designated as responses (i.e., $\Delta F/F_0 > 5$ times the standard deviation above baseline; see Methods). Ca^{2+} transients are seen after γEVG stimulation (orange shaded area) in both neurons. However, only the neuron on the left is confirmed to be γEVG-responsive. The neuron on the right responded equally to the control stimulus (artificial saliva, blue shaded area) as to γEVG, likely reflecting responses to temperature or flow (which always appear as single, transient responses with brief latencies; see Fig. 9.7 and Leijon et al. 2019). Calibs, 30 s, 0.5$\Delta F/F_0$

9.3 Results

9.3.1 Kokumi *Peptide γ-Glu-Val-Gly*

9.3.1.1 Trigeminal Primary Afferent Neurons Respond to the *Kokumi* Peptide γ-Glu-Val-Gly

To investigate if oral exposure to γ-Glu-Val-Gly (γEVG) elicits trigeminal responses in vivo, the oral cavity of the mouse was superfused with 100 μM γEVG. Because many trigeminal ganglion neurons are also sensitive to other sensory stimuli, for example, temperature (Yarmolinsky et al. 2016; Leijon et al. 2019), γEVG was preceded with a control stimulus consisting of artificial saliva at the same temperature and perfusion rate as the γEVG application (Fig. 9.3). Our results show that γEVG (100 μM) applied orally elicits responses in a small population of trigeminal ganglion neurons (Fig. 9.4).

9.3.1.2 Characteristics of Responses Evoked by γ-Glu-Val-Gly

γEVG responses in trigeminal neurons had variable latencies. Although occasionally we observed well-defined, transient γEVG-evoked responses within a few seconds of the stimulus onset (Fig. 9.5a), more often γEVG responses had much longer latencies and were characterized by prolonged bursts of activity (Figs. 9.4 and 9.5b–c). Across a sample of 5 mice (26 cells), the mean latency of γEVG responses was 78 ± 12 s (Fig. 9.6a). The average amplitude was 1.4 ± 0.2 $\Delta F/F_0$ (Fig. 9.6b).

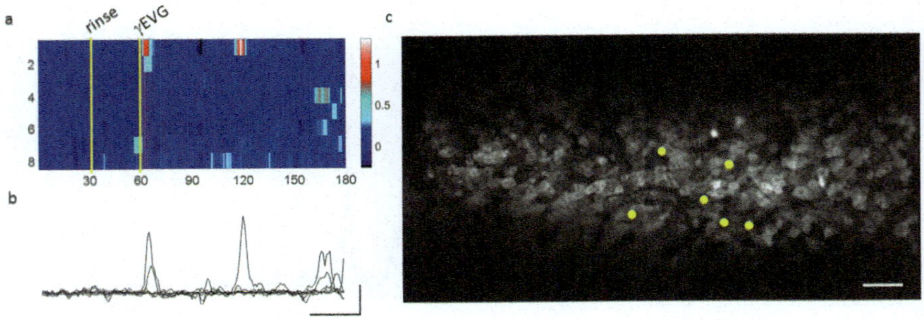

Fig. 9.4 Oral γEVG evokes responses in trigeminal ganglion neurons. (**a**) A heat map displaying the fluorescence change ($\Delta F/F_0$; color bar) of 8 trigeminal ganglion neurons over a 180-s recording. Yellow lines indicate onset of stimuli (stimulus duration = 10 s). The top 6 neurons show responses to 100 μM γEVG. Note that these neurons are silent during the baseline (first 30 s) and are unaffected by the control (artificial saliva) stimulus. The bottom 2 cells show nonspecific activity and are not considered γEVG-responsive. (**b**) Superimposed traces of the 6 γEVG-responsive neurons from recording shown in a. Calibs, 30 s, 0.5 $\Delta F/F_0$. (**c**) Location of the 6 γEVG-responsive neurons in the trigeminal ganglion. Note that imaging the ganglion allows simultaneous recording of a large number of trigeminal ganglion neurons (typically around 1000). Scale bar, 100 μm

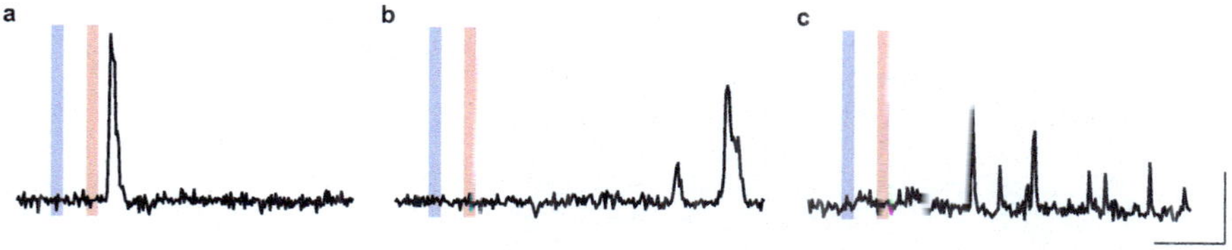

Fig. 9.5 Oral γEVG evokes short- and long-latency responses in trigeminal ganglion neurons. Traces show evoked responses in 3 different neurons to 100 µM γEVG applied orally for 10 s (orange shaded areas). Responses consisted of a single transient (**a**) or multiple transients (**b**, **c**). Note that an initial rinse with artificial saliva for 10 s (blue shaded areas) did not elicit responses in any of these neurons. Calibs, 60 s, 1 $\Delta F/F_0$

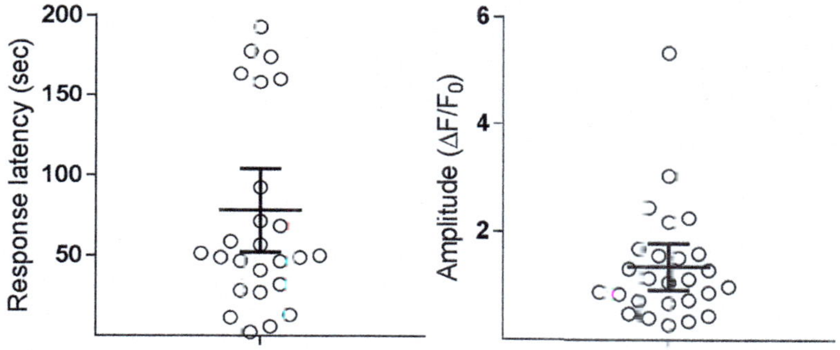

Fig. 9.6 Properties of γEVG-evoked responses. The response latencies (sec) and amplitudes (max $\Delta F/F_0$) were measured across a sample of 5 mice (26 cells). Error bars show mean ± 95% CI

9.3.1.3 Comparing γ-Glu-Val-Gly Responses to Thermal Responses

The incidence of trigeminal ganglion neurons stimulated by oral application of γEVG was quite low (0.6 ± 0.0%, 3 mice). By comparison, the incidence of trigeminal ganglion neurons that responded to orostimulation with cooled artificial saliva (~15 °C) was >10 fold higher, 7.4 ± 0.6%, a significant difference (3 mice; $p < 0.0001$; Fig. 9.7b). From the same sample, 36% (8/22) of γEVG-responsive neurons were also sensitive to cooling of the oral cavity.

The trigeminal ganglion consists of a diverse population of sensory neurons (e.g., mechanosensitive, thermosensitive, nociceptive, itch) with varying cell sizes. The mean cell diameter (Fig. 9.7c) of γEVG-responding neurons was 16.7 ± 0.61 µm, which is significantly smaller than the average of all trigeminal neurons (23.9 ± 0.63, $p < 0.0001$). Cool-sensing neurons, as well, had small diameters (19.1 ± 0.51 µm).

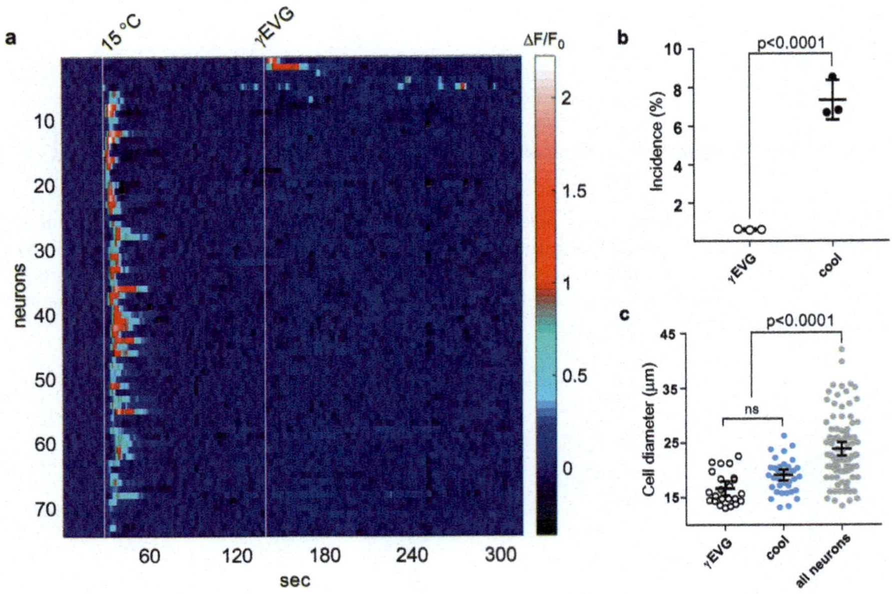

Fig. 9.7 Comparing γEVG responses to responses evoked by cooling the oral cavity. (**a**) Heat map showing the fluorescence change ($\Delta F/F_0$; color bar) of 76 trigeminal ganglion neurons during 10-s oral stimulation (onset at white lines) with cold (~15 °C) artificial saliva and with 100 µM γEVG, sequentially. (**b**) Incidences for γEVG-responding and cool-sensitive neurons (3 mice, 22 and 272 neurons, respectively; Students t-test). (**c**) Cell sizes were measured for γEVG-responsive (7 mice, 23 neurons) and cool-sensitive neurons (3 mice, 36 neurons), as well as for a random sampling of all trigeminal neurons (3 mice, 94 neurons). Error bars show mean ± 95% CI

9.3.1.4 Effect of CaSR Blocker NPS-2143 on γ-Glu-Val-Gly-Evoked *Kokumi* Responses

CaSR is thought to be involved in responses evoked by *kokumi* substances observed (a) in HEK-293 cells transiently expressing human CaSR (Ohsu et al. 2010), (b) in CaSR-expressing taste cells in lingual slices (Maruyama et al. 2012), and (c) in human sensory analyses (Kuroda and Miyamura 2015). Thus, we investigated whether the CaSR inhibitor NPS-2143 had an effect on γEVG-evoked responses recorded in vivo from trigeminal ganglion neurons. We recorded responses to oral stimulation with γEVG alone or in combination with 30 µM NPS-2143 (Fig. 9.8). Because γEVG often elicited multiple calcium transients (Fig. 9.5), we calculated the area under curve (AUC) to quantify γEVG responses. The results showed that NPS significantly decreased γEVG responses ($p < 0.0001$) (Fig. 9.8). NPS had no effect on stimulation with artificial saliva ($p > 0.05$, Fig. 9.8a). To exclude the possibility that the decline in γEVG responses following NPS treatment was merely due to "rundown" during prolonged recordings, we conducted a second series of experiments where γEVG alone was applied twice in the absence of the CaSR inhibitor (Fig. 9.8b). Importantly, there was no decrease in AUC of γEVG responses

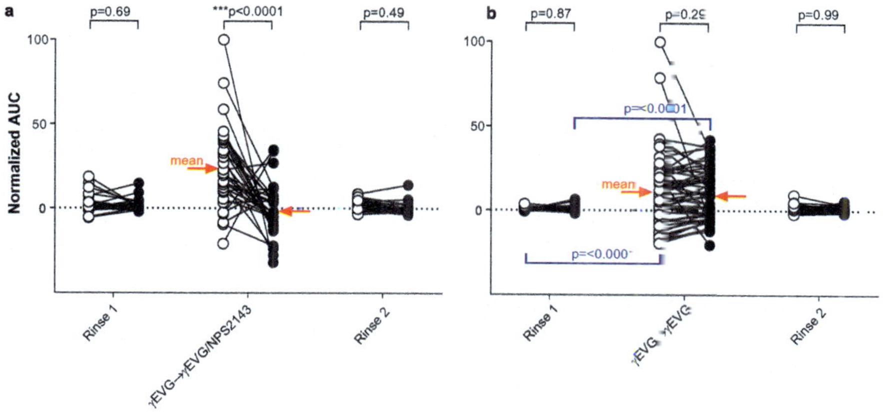

Fig. 9.8 The CaSR inhibitor NPS 2143 reduces γEVG responses in trigeminal ganglion neurons. (**a**) A sequential series of oral stimuli: artificial saliva (rinse 1) → 100 μM γEVG → artificial saliva (rinse 2) was applied and trigeminal neuron responses (area under the curve of $\Delta F/F_0$, AUC) were recorded. After this control series was completed (open symbols), the oral cavity was flushed with 30 μM NPS 2143 and the same sequence of stimuli was repeated (closed symbols). The response to γEVG was significantly depressed in the presence of NPS 2143 (closed symbols; paired t-test). Responses to artificial saliva, although small, were unaffected by NPS 2143 (2 mice, 31 neurons). (**b**) In a second series of experiments, the same protocol was repeated but without NPS 2143 to test for rundown of γEVG responses during the prolonged recording and repeated stimulations. No rundown of γEVG responses occurred (3 mice, 51 neurons)

($p = 0.29$). Furthermore, there was no difference between AUC measurements from γEVG alone responses between the first and second of these series of experiments ($p = 0.5589$). These data suggest that γEVG-evoked responses in the trigeminal ganglion neurons can be inhibited by applying the CaSR inhibitor NPS-2143 into the oral cavity.

9.3.2 *Effect of Viscosity on γ-Glu-Val-Gly-Evoked Responses*

9.3.2.1 γ-Glu-Val-Gly and Xanthan Gum

In human sensory analyses, *kokumi* substances stimulate a sensation of enhanced thickness and mouthfeel during food intake (Ohsu et al. 2010). We investigated whether a viscous solution evoked neuronal responses and whether viscosity affected γEVG responses. Xanthan gum is a common commercial food thickener. Xanthan gum alone (0.5%) appeared to evoke responses in trigeminal ganglion neurons but the effects were diffuse and difficult to quantify (Fig. 9.9). To analyze the data, we removed data ($\Delta F/F_0$) that did not reach a level of five times the standard deviation above baseline and calculated the area under the remaining signal (AUC). We also counted the number of peaks in $\Delta F/F_0$ (Fig. 9.9c). The results show that

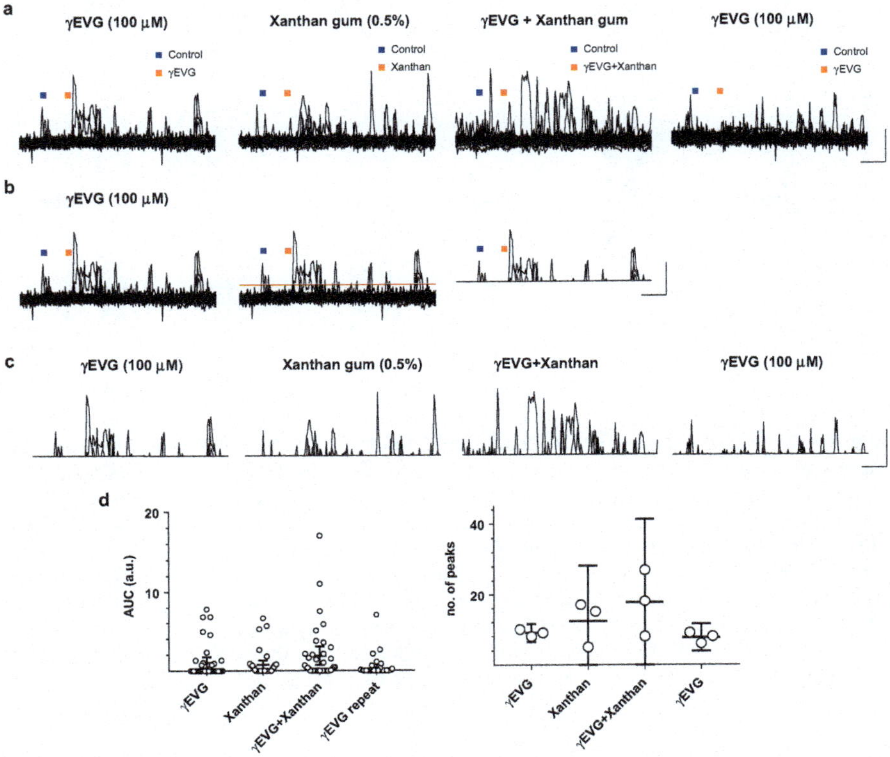

Fig. 9.9 The effect of the food thickener xanthan gum on γEVG-evoked responses. (**a**) Original recordings during oral application of γEVG and xanthan gum individually, as well as combined. (**b**) As exemplified here for a response to γEVG (left trace), for each recording we calculated the signal criterion (equal to >5 s.d. above baseline, red line in middle trace) and eliminated any points less than this value (right trace). (**c**) The responses that reached the cutoff criterion were analyzed for area under the curve (AUC) as well as number of peak values of $\Delta F/F_0$. (**d**) Left, each data point corresponds to AUC for responses noted on the x axis (38 neurons, 3 mice). Right, each data point corresponds to the total number of peaks in $\Delta F/F_0$ traces, per mouse (3 mice). Error bars show mean + 95% CI. Calibs, 30 s, 1 $\Delta F/F_0$

oral xanthan gum alone evoked increased activity in trigeminal ganglion neurons. Stimulating with a combination of γEVG and xanthan gum resulted in a somewhat larger AUC for the combination response than that evoked by γEVG alone, but the difference was not statistically significant ($p = 0.51$). The same result was obtained for the number of peaks ($\Delta F/F_0$) ($p = 0.56$). Thus, oral application of the food thickener xanthan gum can evoke responses in the trigeminal ganglion, but an interaction between xanthan gum and the *kokumi* substance γEVG cannot be confirmed.

9.3.2.2 γ-Glu-Val-Gly and Glucomannan

Next, we tested if another thickener, glucomannan, had an effect on γEVG-evoked responses. Similar to xanthan gum, oral stimulation with glucomannan alone (0.3%) evoked responses in trigeminal ganglion neurons (Fig. 9.10a). There was no interaction between γEVG and glucomannan. Stimulating the oral cavity with a combination of γEVG and glucomannan did not evoke responses that differed from stimulating with γEVG alone (no difference in AUC of the responses, $p = 0.37$, or in number of peaks, $p = 0.65$) (Fig. 9.10b).

9.3.2.3 Sizes of γEVG-Responding Neurons Versus Neurons Responding to Thickeners

In the thickener experiments, some neurons showed responses to both γEVG and thickener (xanthan gum/glucomannan). The average diameter of γEVG-responding neurons was 18.2 ± 0.7 μm (Fig. 9.11), which is not significantly different from our previous measurements (Fig. 9.7, $p = 0.13$). The average sizes of neurons responding to xanthan gum (17.2 ± 1.6 μm) or glucomannan (16.1 ± 0.5 μm) were similar.

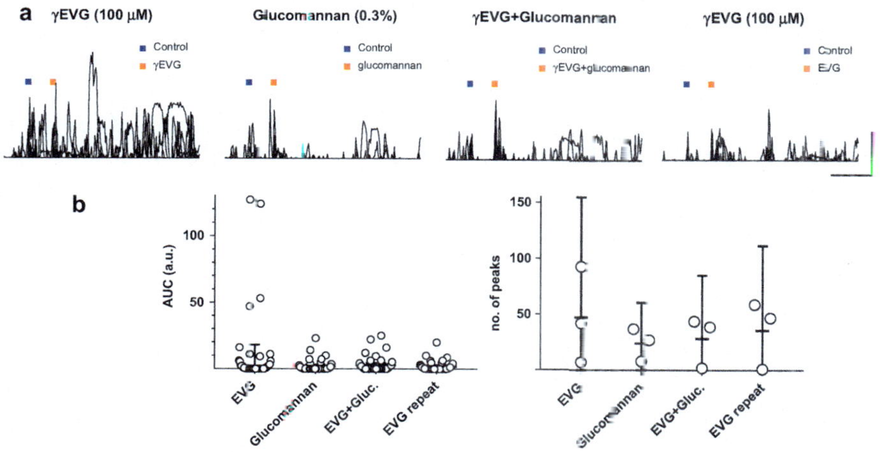

Fig. 9.10 The effect of glucomannan on γEVG-evoked responses. (**a**) As for xanthan gum in Fig. 9.9, responses above the cutoff criterion were analyzed for AUC as well as number of peaks in the trace. (**b**) Left, each data point corresponds to AUC (45 neurons, 3 mice). Right, each data point corresponds to the total number of peaks in the responses, per mouse (3 mice). Responses to glucomannan and γEVG alone and in combination did not differ in either AUC or number of peak responses. Error bars are mean ± 95% CI. Calibs, 50 s, 0.5 $\Delta F/F_0$

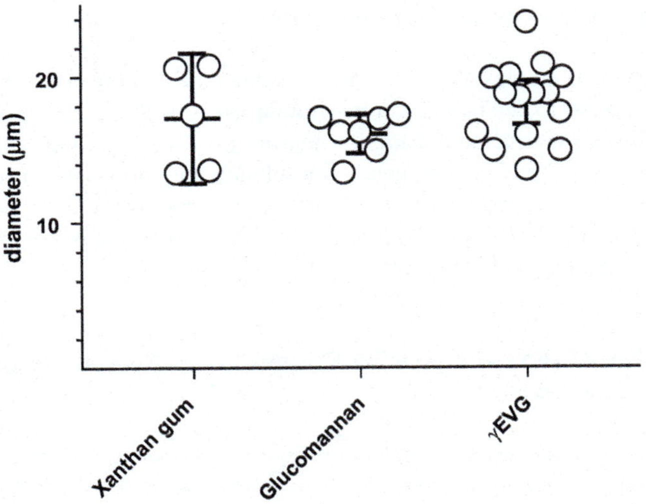

Fig. 9.11 The cell diameter (µm) of neurons that respond to viscous stimuli or to γEVG. There was no difference in the diameters between trigeminal ganglion neurons responding to oral application of xanthan gum (3 mice, 5 cells), glucomannan (3 mice, 7 neurons), or γEVG (3 mice, 17 neurons). Error bars are means ± 95% CI

9.3.3 Additional Kokumi *Substances*

Although γEVG has been described as a potent *kokumi* substance (Kuroda and Miyamura 2015), there are other peptides with *kokumi*-like activity (Amino et al. 2016; Kuroda and Miyamura 2015). We investigated if *kokumi* substances other than γEVG elicited Ca^{2+} responses in mouse trigeminal ganglion neurons. Specifically, we tested glutathione (GSH/γ-Glu-Cys-Gly; 4 mice), γ-Glu-Abu (2 mice), poly-L-lysine (2 mice), $CaCl_2$ (2 mice), and the calcium-mimetic drug cinacalcet (3 mice).

Glutathione (GSH) The effect of GSH on taste has been known since the early 1990s when it was first extracted from garlic water and shown to enhance umami intensity (Ueda et al. 1990). This "intensifying" effect was termed *koku*. GSH has since been shown to also affect other taste qualities (e.g., Ohsu et al. 2010). We found that oral application of 1 mM GSH evoked responses in mouse trigeminal ganglion neurons. The responses (Fig. 9.12a) were of similar latency and duration as those of γEVG (Fig. 9.6). Across 4 mice, GSH evoked responses in 26 trigeminal ganglion neurons.

Cinacalcet Cinacalcet acts as a calcimimetic and causes allosteric activation of CaSR. In human sensory analyses, oral 38 µM cinacalcet showed significant *kokumi*-like taste enhancement (Ohsu et al. 2010). We found that 100 µM cinacalcet evoked responses in the trigeminal ganglion (Fig. 9.12b). Across 3 mice, cinacalcet evoked responses in 12 trigeminal ganglion neurons.

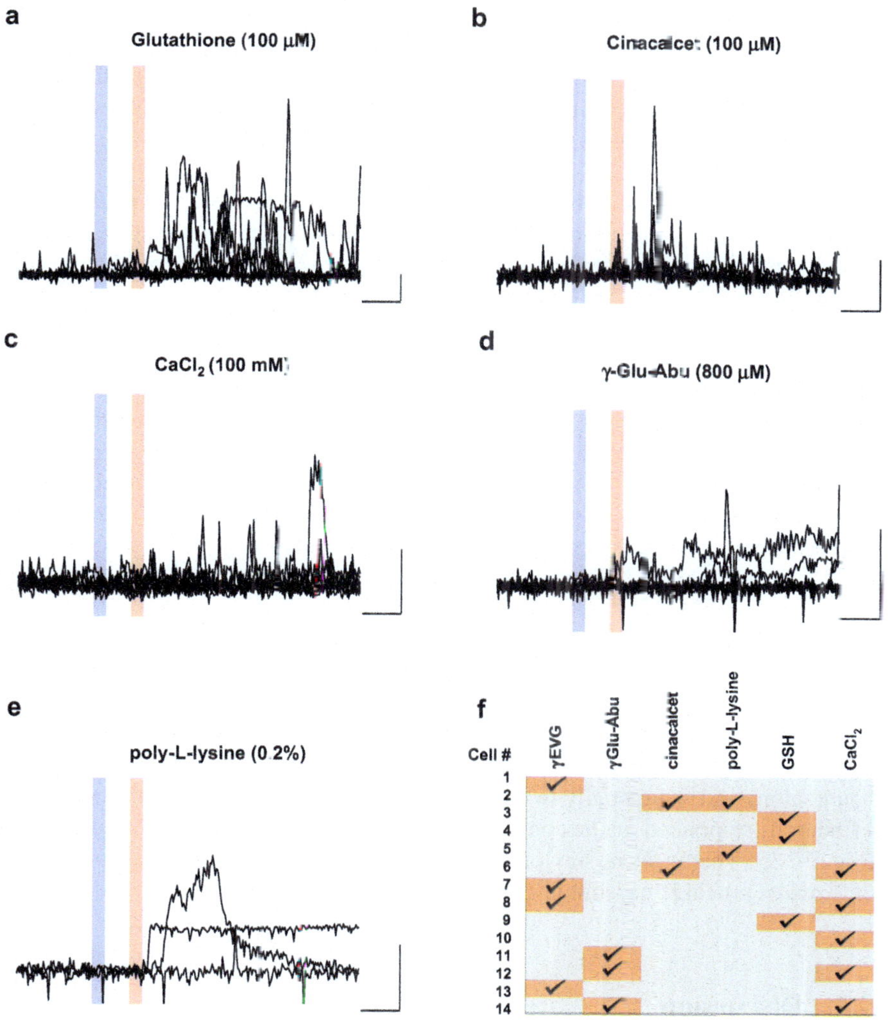

Fig. 9.12 In vivo trigeminal ganglion neuron responses elicited by 5 selected *kokumi* substances applied orally. In (**a**–**e**), each graph contains data from one mouse for which all responding neurons to the respective substance have been overlaid. Shaded areas correspond to the stimuli onsets of artificial saliva (blue) and the respective *kokumi* substance (orange). Note that the control stimulus (artificial saliva) did not elicit responses. (**f**) One experiment was conducted where all 6 *kokumi* substances were applied sequentially. A total of 14 neurons (out of >1000) responded to at least one *kokumi* substance, and 5 of those 14 neurons responded to more than one *kokumi* substance. Calibs, 30 s, 1 $\Delta F/F_0$

CaCl₂ A certain basal level of calcium is needed for *kokumi* substances to activate CaSR (Wang et al. 2006; Ohsu et al. 2010), and calcium salts have *kokumi*-like activity (Ohsu et al. 2010). We superfused the oral cavity with 100 mM $CaCl_2$ and observed trigeminal neuron responses that were similar in delay and characteristics to those of γEVG (Fig. 9.12c). Across 2 mice, $CaCl_2$ evoked responses in 11 neurons.

γ-Glu-Abu The *kokumi* substance γ-glutamyl-α-aminobutyrate (γ-Glu-Abu) is a natural substrate of glutathione synthetase, also known under its systematic name butanoic acid. It is one of the most potent dipeptide *kokumi* substances (Amino et al. 2016). Across 2 mice, γ-Glu-Abu (800 μM) evoked responses in 11 neurons (Fig. 9.12d).

Poly-L-Lysine Polylysine is a basic peptide that activates CaSR (Brown and MacLeod 2001). In human sensory analyses, 0.08% polylysine has been shown sufficient to enhance taste (Ohsu et al. 2010). We found that, in the mouse, oral stimulation with polylysine (0.2%) elicited responses only in a few trigeminal neurons. Polylysine responses were quite prolonged (Fig. 9.12e). Across 2 mice, a total of 6 neurons were responsive to polylysine at this concentration.

9.3.4 *Overlap Between* Kokumi *Substances*

To determine if the above *kokumi* substances elicited responses in identical, overlapping, or separate groups of trigeminal ganglion neurons, we applied each compound alone and sequentially in one experiment. The results indicated that 9 of the 14 neurons responded to one compound only; 5 neurons responded to 2 compounds (Fig. 9.12f). There was no obvious pattern in the overlap. No consistent population of "*kokumi* sensing" trigeminal ganglion neurons was evident.

9.4 Discussion

In this study, we investigated responses of mouse trigeminal ganglion neurons to oral stimulation with *kokumi* compounds. We used transgenic mice that express GCaMP in sensory neurons to conduct in vivo imaging and analyze a number of *kokumi* compounds as well as test the effect of blocking CaSR, a putative receptor for *kokumi* compounds.

Trigeminal sensory neuron responses to γEVG (γ-Glu-Val-Gly) had variable latencies (2–>200 s) and were often characterized by prolonged activity. This likely reflects the gradual penetration of the compound into the lingual epithelium where the trigeminal afferent sensory nerve endings are situated. These trigeminal terminals can be quite superficial in the lingual epithelium (Leijon et al. 2016). During sensory evaluations, panelists report that γEVG added to food enhances taste intensity, peaking at 10 s but still present up to 40 s (Ohsu et al. 2010). Those data would be consistent with the present study where oral stimulation with γEVG evoked long-lasting responses in trigeminal ganglion neurons.

CaSR has previously been implicated as a mechanism of action for *kokumi* taste both in vitro (Amino et al. 2016) and in human sensory analyses (Ohsu et al. 2010). Thus, we tested if the CaSR inhibitor NPS-2143 affected γEVG responses. Indeed, co-application of NPS-2143 significantly decreased γEVG-evoked activity. Whether NPS-2143 selectively and specifically depresses only γEVG responses and not other stimulus-evoked activities (e.g., mechanical, thermal, or nociceptive) remains to be tested. Also, we did not test NPS-2143 on responses to other *kokumi* compounds such as glutathione. Moreover, it is unclear to what extent CaSR is expressed in the trigeminal ganglion. RNAseq analyses revealed CaSR expression only in few (<1%) of mouse trigeminal ganglion neurons and only at negligible levels (Nguyen et al. 2017). Yet if CaSR is only expressed in neurons that innervate the oral mucosa, this might explain the low percentage of trigeminal ganglion neurons that express this receptor and the sparsity of neurons that respond to γEVG (0.6%). It would not, however, explain the low level of CaSR expression per cell that Nguyen et al. (2017) report. Taken collectively, our findings may be suggestive, but do not offer conclusive evidence for the involvement of CaSR in trigeminal neuron responses to *kokumi* substances.

Based on the *koku* characteristics "mouthfulness" or "thick flavor," we studied whether there is a sensory interaction between *kokumi* compounds and food thickeners. The data did not reveal significant interactions between oral γEVG and xanthan gum or glucomannan. Because a population of trigeminal sensory ganglion neurons responded to oral stimulation with γEVG or to xanthan gum (or glucomannan) alone, the perception of *koku*-enhanced thickness that sensory panels report (Ohsu et al. 2010) may be generated by *kokumi* compounds stimulating viscosity-sensitive trigeminal sensory afferent fibers.

The list of compounds identified as *kokumi* substances is growing (e.g., Amino et al. 2016). We tested 5 of these, including γEVG, glutathione (GSH), cinacalcet γ-Glu-Abu, and $CaCl_2$. All these compounds, applied orally, evoked responses in some trigeminal ganglion neurons. The response properties were mostly small (low amplitude), with variable latencies and prolonged durations. However, when tested sequentially (admittedly in a limited sample of neurons), the above five compounds did not stimulate a singular, coherent population of trigeminal ganglion neurons. This suggests that if *koku* sensations are indeed transmitted via somatosensory afferent fibers of the trigeminal ganglion, there is not a uniform group of "*koku*-responding" neurons in that ganglion. Moreover, the relatively small population of trigeminal ganglion neurons dedicated to oral stimulation with *kokumi* substances (for instance, compared with thermal-orosensitive neurons, Leijon et al. 2019) suggest that it will remain a challenge to study *koku*-evoked somatosensations at the cellular level in the trigeminal ganglion. In particular, relating *koku*-responsive neurons to molecular markers and eventually identifying a molecular receptor in trigeminal neurons will be challenging.

It is worth noting that *kokumi* substances also have been shown to activate CaSR-expressing taste bud cells (Maruyama et al. 2012). These cells are innervated by afferent fibers from the geniculate ganglion. Indeed, taste buds in the tongue and palate are associated with fibers from both geniculate and trigeminal ganglia (Fig. 9.13). Neurons from the geniculate ganglion innervate taste bud cells that are stimulated by taste substances entering through an apical taste pore. Thus, geniculate ganglion neurons typically are thought to transmit basic taste qualities such as sweet, umami, etc. In contrast, axons from trigeminal ganglia penetrate the oral epithelium immediately surrounding taste buds, forming a halo of nerves around most taste buds. Importantly, trigeminal fibers frequently are located in very superficial layers of the epithelium, where they may be reached by oral stimuli such as γEVG. In summary, trigeminal and geniculate ganglion neurons are positioned to respond to oral stimulation with *kokumi* substances—trigeminal neurons by direct stimulation and geniculate neurons following stimulation of taste bud cells. The sensory characteristics of *kokumi* compounds are diverse, and it is thus possible that the trigeminal and geniculate ganglia play different, or complimentary, roles in the multifaceted taste that is *koku*.

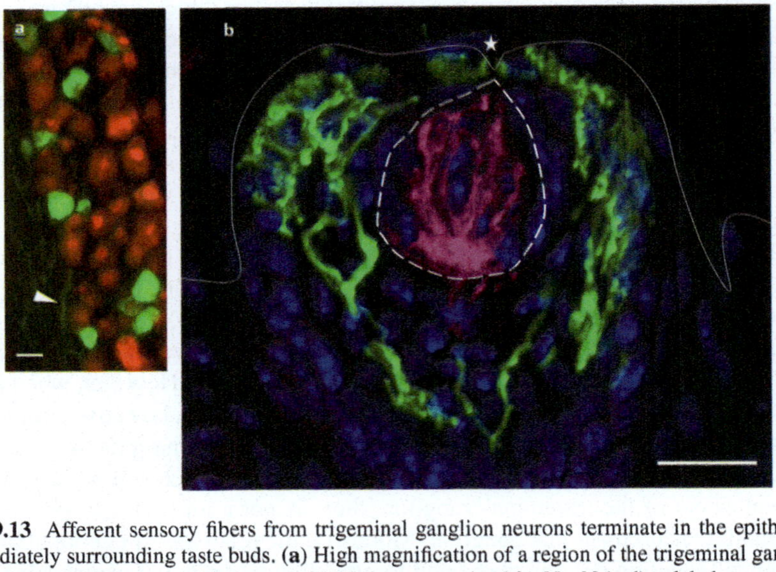

Fig. 9.13 Afferent sensory fibers from trigeminal ganglion neurons terminate in the epithelium immediately surrounding taste buds. (**a**) High magnification of a region of the trigeminal ganglion from a *Trpm8*-GFP mouse (Dhaka et al. 2007), immunostained for NeuN (red) to label neurons and GFP (green) to mark those neurons that express TrpM8. A subset of the neurons expresses GFP/TrpM8. GFP is also present in axons from these neurons (arrowhead). The geniculate ganglia from the same mouse, processed in parallel, did not contain GFP/TrpM8 (not shown). (**b**) Confocal micrograph of a fungiform papilla (solid outline) with a taste bud (dashed outline) in the tongue of the same *Trpm8*-GFP mouse. GFP/TrpM8-expressing axons surround, but do not enter, the taste bud. In contrast, afferent fibers from the geniculate (gustatory) ganglion are immunoreactive for P2X2 (magenta) and branch extensively throughout the taste bud. The taste pore, through which taste stimuli interact with taste bud cells, is indicated (∗). Scale bars, 20 μm

Acknowledgments This work was supported by grants from Ajinomoto Co., Inc., and the USA National Institutes of Health: NIDCD R01DC014420 (SR, NC) and NIDCR/NCI R21DE027237 (SR). The authors thank Dr. Ajay Dhaka, University of Washington, for the generous donation of tissue used to produce Fig. 9.13.

References

Amino Y, Nakazawa M, Kaneko M, Miyaki T, Miyamura N, Maruyama Y Eto Y (2016) Structure–CaSR–activity relation of *Kokumi* γ-glutamyl peptides. Chem Farm Bull 64(8):1181–1189

Brown EM, MacLeod RJ (2001) Extracellular calcium sensing and extracellular calcium signaling. Physiol Rev 81:239–297

Dhaka A, Murray AN, Mathur J, Earley TJ, Petrus MJ, Patapoutian A (2007) TRPM8 is required for cold sensation in mice. Neuron 54:371–378

Dunkel A, Koster J, Hofmann T (2007) Molecular and sensory characterization of gamma-glutamyl peptides as key contributors to the *kokumi* taste of edible beans (Phaseolus vulgaris L.). J Agric Food Chem 55:5712–6719

Heyeraas KJ, Haug SR, Bukoski RD et al (2008) Identification of a Ca^{2+}-sensing receptor in rat trigeminal ganglia, sensory axons, and tooth dental pulp. Calcif Tissue Int 82:57

Kim YS, Chu Y, Han L, Li M, Li Z, LaVinka PC, Sun S et al (2014) Central terminal sensitization of TRPV1 by descending serotonergic facilitation modulates chronic pain. Neuron 81(4):873–887

Kuroda M, Miyamura N (2015) Mechanism of the perception of "*kokumi*" substances and the sensory characteristics of the "*kokumi*" peptide, γ-Glu-Val-Gly. Flavor 4:11

Leijon, S, Breza, JM, Berger, M, Maruyama, Y, Chaudhari N, Roper SD (2016) In vivo imaging of trigeminal ganglion neuron responses to γEVG, capsaicin, allyl isothiocyanate (AITC), and menthol in mice. Abstract presented at ISOT 2016, Yokohama, Japan

Leijon SCM, Neves AF, Breza JM, Simon SA, Chaudhari N, Roper SD (2019) Oral thermosensing by murine trigeminal neurons: modulation by capsaicin, menthol, and mustard oil. J Physiol. In press 597:2045

Maruyama Y, Yasuda R, Kuroda M, Eto Y (2012) *Kokumi*. Substance, enhancers of basic tastes, induce responses in calcium-sensing receptor expressing taste cells. PLoS One 7(4):e34489. https://doi.org/10.1371/journal.pone.0034489

Nguyen MQ, Wu Y, Bonilla LS, von Buchholtz LJ, Ryba NJP (2017) Diversity amongst trigeminal neurons revealed by high throughput single cell sequencing. PLoS One 12(9):e0185543

Ohsu T, Amino Y, Nagasaki H, Yamanaka T, Takeshita S, Hatanaka MY, Miyamura N, Eto Y (2010) Involvement of the calcium-sensing receptor in human taste perception. J Biol Chem 285(2):1016–1022

Sollars SI, Hill DL (2005) In vivo recordings from rat geniculate ganglia: taste response properties of individual greater superficial petrosal and chorda tympani neurones. J Physiol 564(3):877–893

Ueda Y, Sakaguchi M, Hirayama K, Miyajima R, Kimizuka A (1990) Characteristic flavor constituents in water extract of garlic. Agric Biol Chem 54:163–169

Wang M, Yao Y, Kuang D, Hampson D (2006) Activation of family C G-protein-coupled receptors by the tripeptide glutathione. J Biol Chem 281:8864–8870

Wu A, Dvoryanchikov G, Pereira E, Chaudhari N, Roper SD (2015) Breadth of tuning in taste afferent neurons varies with stimulus strength. Nat Commun 6:8171

Yarmolinsky DA, Peng Y, Pogorzala LA, Rutlin M, Hoon MA, Zuker CS (2016) Coding and plasticity in the mammalian thermosensory system. Neuron 92(5):1079–1092

Chapter 10
Overview and Future Prospect of Studies on *Koku*

Toshihide Nishimura and Motonaka Kuroda

Abstract In this chapter, the overview of studies on *koku* is described. The definition of *koku* is proposed as a sensation such as complexity, mouthfulness, and lingeringness, and mainly given by lots of stimulations by taste compounds, aroma compounds, and texture. In this book, the research topics on the effects and action mechanisms to form and/or enhance sensation in *koku* attributes were described. These compounds should be called *koku*-imparting substances. Although the receptor for several *koku*-imparting substances such as MSG, *kokumi* γ-glutamyl peptides, and aromatic compounds has been identified, the action mechanism of *koku* impartment by these compounds remains unclear. Further scientific studies in the field of food science (sensory science, food chemistry, and food processing), physiology, neurobiology, molecular biology, and brain science are needed.

Keywords *Koku* · Taste · Smell · Odor · Texture · Tactile · Mechanism

10.1 Definition of *Koku*

Figure 1 in Chap. 1 has shown that there are lots of objective factors derived from foodstuffs contributing to palatability of foods. We propose that *koku* is also an objective factor involved in food palatability as follows. We also propose that *koku* is a sensation such as complexity, mouthfulness, and lingeringness, which is mainly given by lots of stimulations in taste, aroma, and texture. Therefore, components stimulating smell sensation, tactile sensation, and taste sensation can enhance *koku* in foods. In authentic foods, aroma, taste, and texture components produced by heating, fermentation, and conditioning can form the *koku* sensation. The research on *koku* attribute has just started, and some compounds are shown to be involved in

T. Nishimura (✉)
Faculty of Nutrition, Kagawa Nutrition University, Sakado, Saitama, Japan
e-mail: nishimura.toshihide@eiyo.ac.jp

M. Kuroda
Institute of Food Sciences & Technologies, Ajinomoto,
Suzuki-cho, Kawasaki, Kanagawa, Japan

T. Nishimura, M. Kuroda (eds.), *Koku in Food Science and Physiology*,
https://doi.org/10.1007/978-981-13-8453-0_10

the formation or enhancement in *koku* sensation. In several chapters, the research topics on the effects and action mechanisms to form and/or enhance sensation in *koku* attributes were described. Chapters 2 and 4 provide that umami substances can enhance the sensation mouthfulness and lingeringness in *koku* attribute of foods. Especially, in Chap. 2, Nishimura et al. (2016a, b) have shown that these substances enhance the intensity of retronasal aroma after putting foods into oral cavity. In Chap. 4, edible oils also enhance the sensations complexity and continuity by their capacity-holding aroma compounds in pork sausages (Nishimura et al. 2016b). Chapter 5 has shown that aromatic components such as phthalides, tridecatrienal, and rotundones can enhance mouthfulness by their addition to foods at the lower concentration than threshold level. In Chap. 6, aromatic compounds in beer give complexity in *koku* attributes and are produced by the fermentation in its production process. In Chaps. 7, 8, and 9, *kokumi* substances such as glutathione and γ-Glu-Val-Gly are shown to have enhancing effect on mouthfulness and lingeringness of foods by addition at the lower concentration than threshold level.

10.2 Taste-Related *Koku*-Enhancing Substances

For taste-related *koku*-enhancing factors, the effects of monosodium glutamate (MSG) on sensory characteristics related to *koku* are described in Chaps. 2 and 4. In Chap. 2, Yamamoto indicated that glutamate (MSG) has a dual function; MSG has a unique basic taste which is not perceived delicious; however, it makes food delicious when added to food. He cited research conducted by Yamaguchi and Komizuka (1979), who showed that the taste of MSG is neutral or not palatable by itself, but when MSG was added to beef consommé, the flavor characteristics such as thickness, continuity, mouthfulness, impact, and mildness were increased. Many of these characteristics were considered as related to the *koku* attribute (described in Chap. 1). Chapter 4 describes that addition of MSG to sausage enhanced the flavor characteristics such as complexity and lingeringness. These characteristics were also considered as related to the *koku* attribute. Based on these results, umami substances such as MSG are thought to be *koku*-enhancing substances. Potent *kokumi* substances, γ-Glu-Val-Gly and glutathione, were also proposed to be taste-related *koku*-enhancing substances. As shown in Chap. 7, *kokumi* substances modify the basic taste and lingerlingness, mouthfulness, and thickness of taste, when they are added at the lower concentration than threshold level to a basic taste solution or some foods such as chicken consommé, reduced-fat peanut butter, reduced-fat custard cream, and reduced-fat French dressing. Because various sensory characteristics enhanced by γ-Glu-Val-Gly are related to the *koku* attribute, *kokumi* substances are also considered as taste-related *koku*-enhancing substances.

10.3 Odor-Related *Koku*-Enhancing Substances

In Chaps. 5 and 6, studies of odor-related *koku*-enhancing substances were described. In Chap. 6, it was shown that the addition at subthreshold levels of phthalides found in celery increased the intensity of thick, mild, lasting, and complex sensation when added to chicken broth. Further, addition of 5 ppb (*4Z*, *7Z*)-trideca-4,7-dienal found in dried bonito significantly increased the intensity of thickness when added to a dried bonito-flavored model solution. Moreover, adding rotundone at the subthreshold levels increased the intensity of complex sensation when added to several types of model fruit juice. In Chap. 5, the aroma components that contribute to the thickness of beer aroma were investigated. The results show that 76 aroma compounds including 27 compounds present at lower concentrations than the recognition threshold concentration are necessary to reproduce the thickness of beer aroma. These results suggest that several aroma compounds function as *koku*-enhancing substances at concentrations lower than the recognition threshold in foods.

In Chap. 4, the contribution of oil to *koku* in foods was demonstrated. In sausage, addition of fats enhanced the lingeringness by increasing the retention and sustained release of aroma components. Additionally, Nishimura et al. (2016a, b) reported that phytosterols in onion contribute to the lingering aroma by retaining aroma compounds such as disulfides. These results also suggest that aroma components contribute to complexity in *koku* attributes in foods.

10.4 Texture-Related *Koku*-Imparting Substances

Although studies on texture-related *koku*-enhancing substances are not shown in the book, it has been reported that several texture-modifying substances can enhance the *koku*-related sensory characteristics in foods. As mentioned in Chap. 7, Tomaschunas et al. (2013) reported that addition of inulin and citrus fiber, kinds of texture modifiers, to low-fat Lyon-style sausage increased the intensity of aftertaste meat flavor. Additionally, Liou and Grun (2007) reported that addition of fat mimetics such as microparticulated whey protein concentrate, one of texture-modifying agents, and polydextrose powder enhanced thickness. These previous studies suggest that texture-related substances also contribute to *koku* attribute in foods.

Furthermore, several studies indicated that taste compounds also stimulate the trigeminal sensations. Oka et al. (2013) suggested the involvement of TRP channels located in trigeminal nerves in the perception of salty taste at higher concentrations. Beauchamp (2009) indicated that MSG, a representative umami compound, also has an apparent tactile or feeling component often described as "fullness" as well as classical taste. Additionally, in Chap. 7, it was shown that addition of γ-Glu-Val-Gly, one of *kokumi* substances, significantly increased the intensity of "mouth-coating" and tended to enhance viscous sensation. These observations suggest that taste-related compounds can affect the tactile sensation; however, the mechanism of these actions remains unclear.

10.5 Mechanism of the Perception of *Koku*-Imparting Substances

In this book, the researches of several *koku*-imparting factors are introduced. For taste-related *koku*-imparting factors, umami substances such as MSG are perceived through T1R1/T1R3 heterodimer molecule and mGluRs. The study described in Chap. 8 indicated that the calcium-sensing receptor (CaSR) is involved in the perception of *kokumi* substances. Additionally, the study in Chap. 9 showed that CaSR expression on trigeminal ganglion and mouse trigeminal ganglion respond to *kokumi* substances, suggesting the relationship between the CaSR and the enhancement of texture-related attributes such as mouth-coating sensation and viscous sensation by *kokumi* substances (Chap. 8). For odor-related *koku*-imparting substances, aroma compounds are perceived by olfaction receptors (ORs) located in the olfactory epithelium. Although the receptors for *koku*-imparting substances have been identified, the mechanism of *koku* impartment, that is, the role of receptor in the sensation of *koku* attribute, remains unclear.

10.6 Future Prospect on the Study on *Koku*

The research on *koku* attribute has just started, and only some compounds are shown to be involved in the formation or enhancement in *koku* sensation. The components involved in the sensation complexity, mouthfulness and lingeringness of *koku* attributes are different in foods. The research on compounds in foods possessing *koku* attributes contributing to each sensation is necessarily carried on. Physiological research is also needed. As described above, although the receptors for *koku*-imparting substances such as MSG, *kokumi* substances, and aroma compounds have been identified, the mechanism of *koku* impartment given by these compounds is still unclear. Further scientific studies in the field of food science (sensory science, food chemistry, and food processing), physiology, neurobiology, molecular biology and brain science are needed. The combination of the studies will contribute to further understanding of *koku* in foods.

References

Beauchamp GK (2009) Sensory and receptor responses to umami: an overview of pioneering work. Am J Clin Nutr 90(Suppl):723S–727S

Liou BK, Grun IU (2007) Effect of fat level on the perception of five flavor chemicals in ice cream with or without fat mimetics by using a descriptive test. J Food Sci 72:S595–S604
Nishimura T, Goto S, Miura K, Takakura Y, Egusa AS, Wakabayashi H (2016a) Umami compounds enhance the intensity of retronasal sensation of aromas from model chicken soups. Food Chem 196:577–583
Nishimura T, Egusa AS, Nagao A, Odahara T, Sugise T, Mizoguchi N, Nosho Y (2016b) Phytosterols in onion contribute to a sensation of lingering of aroma, a koku attribute. Food Chem 192:724–728
Oka Y, Butnaru M, von Buchholtz L, Ryba NJP, Zuker CS (2013) High salt recruits aversive taste pathways. Nature 494:472–475
Tomaschunas M, Zorb R, Fischer J, Kohn E, Hinrichs J, Busch-Stockfisch M (2013) Changes in sensory properties and consumer acceptance of reduced fat pork Lyon-style and liver sausages containing inulin and citrus fiber as fat replacers. Meat Sci 95:629–640
Yamaguchi S, Kimizuka A (1979) Psychometric studies on the taste of monosodium glutamate. In: Filer LJ Jr, Garattini S, Kare MR, Reynolds WA, Wurtman RJ (eds) Glutamic acid: advances in biochemistry and physiology. Raven Press, New York, pp 35–54

The manufacturer's authorised representative in the EU is Springer Nature Customer Service Centre GmbH, Europaplatz 3, 69115 Heidelberg, Germany. If you have any concerns regarding our products, please contact ProductSafety@springernature.com

Printed and bound by CPI Group (UK) Ltd, Croydon, CR0 4YY
30/09/2026
02225438-0005